T0242052

Lecture Notes in Computer Science 9240

Commenced Publication in 1973
Founding and Former Series Editors:
Gerhard Goos, Juris Hartmanis, and Jan van Leeuwen

Editorial Board

David Hutchison
  Lancaster University, Lancaster, UK
Takeo Kanade
  Carnegie Mellon University, Pittsburgh, PA, USA
Josef Kittler
  University of Surrey, Guildford, UK
Jon M. Kleinberg
  Cornell University, Ithaca, NY, USA
Friedemann Mattern
  ETH Zurich, Zürich, Switzerland
John C. Mitchell
  Stanford University, Stanford, CA, USA
Moni Naor
  Weizmann Institute of Science, Rehovot, Israel
C. Pandu Rangan
  Indian Institute of Technology, Madras, India
Bernhard Steffen
  TU Dortmund University, Dortmund, Germany
Demetri Terzopoulos
  University of California, Los Angeles, CA, USA
Doug Tygar
  University of California, Berkeley, CA, USA
Gerhard Weikum
  Max Planck Institute for Informatics, Saarbrücken, Germany

More information about this series at http://www.springer.com/series/8851

Ngoc Thanh Nguyen (Ed.)

Transactions on Computational Collective Intelligence XVIII

Springer

Editor-in-Chief

Ngoc Thanh Nguyen
Department of Information Systems
Wroclaw University of Technology
Wroclaw
Poland

ISSN 0302-9743   ISSN 1611-3349   (electronic)
Lecture Notes in Computer Science
ISBN 978-3-662-48144-8   ISBN 978-3-662-48145-5   (eBook)
DOI 10.1007/978-3-662-48145-5

Library of Congress Control Number: 2015945327

Springer Heidelberg New York Dordrecht London
© Springer-Verlag Berlin Heidelberg 2015
This work is subject to copyright. All rights are reserved by the Publisher, whether the whole or part of the
material is concerned, specifically the rights of translation, reprinting, reuse of illustrations, recitation,
broadcasting, reproduction on microfilms or in any other physical way, and transmission or information
storage and retrieval, electronic adaptation, computer software, or by similar or dissimilar methodology now
known or hereafter developed.
The use of general descriptive names, registered names, trademarks, service marks, etc. in this publication
does not imply, even in the absence of a specific statement, that such names are exempt from the relevant
protective laws and regulations and therefore free for general use.
The publisher, the authors and the editors are safe to assume that the advice and information in this book are
believed to be true and accurate at the date of publication. Neither the publisher nor the authors or the editors
give a warranty, express or implied, with respect to the material contained herein or for any errors or
omissions that may have been made.

Printed on acid-free paper

Springer-Verlag GmbH Berlin Heidelberg is part of Springer Science+Business Media
(www.springer.com)

Preface

It is my pleasure to present the XVIII volume of LNCS *Transactions on Computational Collective Intelligence*. This volume includes nine interesting and original papers that have been selected via the peer-review process.

The first paper, "Using Semantic Web for Generating Questions: Do Different Populations Perceive Questions Differently?" by Thinh Le Nguyen, is devoted to the problem of organizing effective learning processes in intelligent e-learning systems. The author proposes an approach to using Semantic Web data for generating questions that are intended to help people develop arguments in a discussion session.

The second paper entitled "Reflection of Intelligent e-Learning/Tutoring — The Flexible Learning Model in LMS Blackboard" by Ivana Simonova, Petra Poulova, and Pavel Kriz describes the theoretical background and practical concept of teaching/learning through online courses as an example of a smart solution of e-learning system. The authors consider personalization based on individual learning preferences, including students' reflection on the individualized online instruction.

In the third paper, "GLIO: A New Method for Grouping Like-Minded Users," Soufiene Jaffali et al. present a novel unsupervised method for grouping like-minded users within social networks. The idea of their method is based on detecting groups of users sharing the same interest centers and having similar opinions. Owing to this it can extract the interest centers and retrieve the polarities from the user's textual posts.

The fourth paper, "A Preferences-Based Approach for Better Comprehension of User Information Needs," by Sondess Missaoui and Rim Faiz desribes a model that can identify which contextual dimensions have a strong influence on the outcome of the retrieval process and should therefore be in the user's focus. In order to achieve these objectives, the authors create a new query language model based on the user's preferences. Next they extend this model in order to define a relevance measure for each contextual dimension for automatically classifying each dimension.

The fifth paper entitled "Performance Evaluation of the Customer Relationship Management Agent's in a Cognitive Integrated Management Support System" by Marcin Hernes concerns essential issues related to the sentiment analysis of customers' opinions performed by customer relationship management agents running in a multi-agent cognitive integrated management information system. This system uses computational collective intelligence methods and allows for supporting the management processes related with all the domains of the enterprise's functioning.

In the sixth paper, "Agreements Technologies — Towards Sophisticated Software Agents in Multi-Agent Environments," Mirjana Ivanović and Zoran Budimac present an approach for using agreement technologies in implementation of sophisticated autonomous software agents that mutually negotiate in order to achieve win–win situations. The authors describe the key concepts in this area and highlight the influence of agreement technologies on the development of more sophisticated multiagent systems.

An overview of several interesting systems and environments from different domains is presented.

The next paper, "Identification of Underestimated and Overestimated Web Pages Using PageRank and Web Usage Mining Methods," by Jozef Kapusta, Michal Munk, and Martin Drlík describes a new method of website analysis and optimization that combines methods of web usage and web structure mining — discovering web users' behavior patterns. It can identify the web pages in which the value of their importance, estimated by the website developers, does not correspond to the real behavior of the website visitors.

In the eighth paper, entitled "Massive Classification with Support Vector Machines," Thanh Nghi Do and Hoai An Le Thi propose an extension of the PSVM, LS-SVM, and NSVM algorithms in several ways to efficiently classify large datasets. They have developed a row-incremental version algorithm for datasets with billions of data points. They also worked out new algorithms to process datasets with a small number of data points but very high dimensionality.

In the last paper, "On a Multi-Agent Distributed Asynchronous Intelligence-Sharing and Learning Framework," Shashi Shekhar Jha and Shivashankar B. Nair develop a framework for realizing distributed and asynchronous sharing of intelligence and consequent learning among the entities of a networked distributed system.

I would like to sincerely thank the authors for their valuable contributions.

June 2015   Ngoc Thanh Nguyen

Transactions on Computational Collective Intelligence

This Springer journal focuses on research in applications of the computer-based methods of computational collective intelligence (CCI) and their applications in a wide range of fields such as the Semanti Web, social networks, and multi-agent systems. It aims to provide a forum for the presentation of scientific research and technological achievements accomplished by the international community.

The topics addressed by this journal include all solutions of real-life problems for which it is necessary to use CCI technologies to achieve effective results. The emphasis of the papers is on novel and original research and technological advancements. Special features on specific topics are welcomed.

Editor-in-Chief

Ngoc Thanh Nguyen   Wroclaw University of Technology, Poland

Co-Editor-in-Chief

Ryszard Kowalczyk   Swinburne University of Technology, Australia

Editorial Board

John Breslin	National University of Ireland, Galway, Ireland
Longbing Cao	University of Technology Sydney, Australia
Shi-Kuo Chang	University of Pittsburgh, USA
Oscar Cordon	European Centre for Soft Computing, Spain
Tzung-Pei Hong	National University of Kaohsiung, Taiwan
Gordan Jezic	University of Zagreb, Croatia
Piotr Jędrzejowicz	Gdynia Maritime University, Poland
Kang-Huyn Jo	University of Ulsan, Korea
Yiannis Kompatsiaris	Centre for Research and Technology Hellas, Greece
Jozef Korbicz	University of Zielona Gora, Poland
Hoai An Le Thi	Lorraine University, France
Pierre Lévy	University of Ottawa, Canada
Tokuro Matsuo	Yamagata University, Japan
Kazumi Nakamatsu	University of Hyogo, Japan
Toyoaki Nishida	Kyoto University, Japan
Manuel Núñez	Universidad Complutense de Madrid, Spain
Julian Padget	University of Bath, UK
Witold Pedrycz	University of Alberta, Canada
Debbie Richards	Macquarie University, Australia
Roman Słowiński	Poznan University of Technology, Poland
Edward Szczerbicki	University of Newcastle, Australia

Tadeusz Szuba   AGH University of Science and Technology, Poland
Kristinn R.   Reykjavik University, Iceland
  Thorisson
Gloria Phillips-Wren   Loyola University Maryland, USA
Sławomir Zadrożny   Institute of Research Systems, PAS, Poland
Bernadetta Maleszka   Wroclaw University of Technology, Poland

Contents

Using Semantic Web for Generating Questions: Do Different Populations Perceive Questions Differently?

Nguyen-Thinh Le[✉]

Department of Informatics Research Group "Computer Science
Education/Computer Science and Society", Humboldt-Universität zu Berlin,
Unter den Linden 6, 10099 Berlin, Germany
Nguyen-thinh.le@hu-berlin.de

Abstract. In this paper, I propose an approach to using semantic web data for generating questions that are intended to help people develop arguments in a discussion session. Applying this approach, a question generation system that exploits WordNet for generating questions for argumentation has been developed. This paper describes a study that investigates a research question of whether different populations perceive questions (either generated by a system or by human experts) differently. To conduct this study, I asked eight human experts of the argumentation and the question generation communities to construct questions for three discussion topics and used a question generation system for generating questions for argumentation. Then, the author invited three groups of researchers to rate the mix of questions: (1) computer scientists, (2) researchers of the argumentation and question generation communities, and (3) student teachers for Computer Science. The evaluation study showed that human-generated questions were perceived differently by three different populations over three quality criteria (the understandability, the relevance, and the usefulness). For system-generated questions, the hypothesis could only be confirmed on the criteria of relevance and usefulness of questions. This contribution of the paper motivates researchers of question generation to deploy various techniques to generate questions adaptively for different target groups.

Keywords: Semantic web · Linked open data · Question generation · Question taxonomy · Adaptivity

1 Introduction

Asking questions is an important skill that is required in many institutional settings, e.g., interviews conducted by journalists (Clayman and Heritage 2002), medical settings (Drew and Heritage 1992), courtrooms (Alkinson and Drew 1979). For teachers, asking questions is almost an indispensable teaching technique. Dillon (1988) investigated questions generated by teachers in 27 upper classrooms in six secondary schools in the USA and reported that questions accounted for over 60 % of the teachers' talk. The benefits of using questions in instruction are multi-faceted and have been reported in many research studies (Lin et al. 2014; Morgan and Saxton 2006;

© Springer-Verlag Berlin Heidelberg 2015
N.T. Nguyen (Ed.): Transactions on CCI XVIII, LNCS 9240, pp. 1–19, 2015.
DOI: 10.1007/978-3-662-48145-5_1

Tenenberg and Murphy 2005). Not only teachers' questions can enhance learning, but also students' question asking can benefit learning. The evidence from research studies provides a solid empirical basis to support the inclusion of students' question asking in teaching in order to enhance comprehension (Rothstein and Santana 2014), cognitive and metacognitive strategies use (Yu and Pan 2014), and problem-solving abilities (Barlow and Cates 2006) of students. Researchers suggested that teachers should pose questions that encourage higher-level thinking of students because they need to be familiarized with different levels of thinking and to use knowledge of the lower-level productively (Chafi and Elkhouzai 2014). In addition, Morgan and Saxton (2006) demonstrated that well-chosen higher-order questions can not only be used to assess student's knowledge but also to extend his/her knowledge, to improve his/her skills of comprehension and application of facts and also to develop his/her higher-order thinking skills. Yet the evidence is that the majority of questions teachers use in their classrooms in order to check knowledge and understanding, to recall of facts or to diagnose student's difficulties (Chafi and Elkhouzai 2014), and only about 10 % of questions are used to encourage students to think (Brown and Wragg 1993). Especially, pre-service teachers, who have just graduated their study, would have many difficulties in generating questions in their classes.

Many automatic question generation approaches have been developed in order to help teachers and students. For example, in the LISTEN[1] project (Mostow and Chen (2009); Mostow and Beck (2007)), Mostow and colleagues developed an automated reading tutor which deploys automatic question generation to improve the comprehension capabilities of students while reading a text. Kunichika et al. (2001) proposed an approach to extracting syntactic and semantic information from an original text and questions are constructed using the extracted information to support novices in learning English. Heilman and Smith (2009) developed an approach to generating questions for assessing students' acquisition of factual knowledge from reading materials. What all these approaches have in common is that they deployed information in a given text (e.g., a reading text) to generate questions.

This paper proposes to use existing encyclopedic or lexical knowledge databases available on the Internet as semantic sources for generating questions automatically. Using the semantic web as a source of information required for generating questions may save time for teachers in preparation for their lessons. Currently, this approach has been experimented by Jouault and Seta (2014) who proposed to generate semantics-based questions by querying information from Wikipedia to facilitate learners' self-directed learning. Using this system, students in self-directed learning are asked to build a timeline of events of a history period with causal relationships between these events given an initial document. The student develops a concept map containing a chronology by selecting concepts and relationships between concepts from a given initial Wikipedia document to deepen their understandings. While the student creates a concept map, the system also integrates the concept to its map and generates its own concept map by referring to semantic information of Wikipedia. The system's concept map is updated with every modification of the student's one and enriched with related

---

[1] http://www.cs.cmu.edu/ ~ ./listen/

concepts that can be extracted from Wikipedia. Thus, the system's concept map always contains more concepts than the student's map. Using these related concepts and their relationships, the system generates questions for the student to lead to a deeper understanding without forcing to follow a fixed path of learning. Also exploiting semantic web data sources, Le et al. (2014) proposed to use WordNet to generate questions that are intended to help students develop arguments for discussion. Their project aimed at using automatically generated questions for stimulating the brain-storming of students during the process of argumentation. WordNet has been used as a semantic source for generating questions, because it is a rich lexical database that is able to provide hyponyms (related concepts) to a queried concept. In a recent study, (Le and Pinkwart 2015) investigated the quality of system-generated questions with respect to the understandability, the relevance of questions to a given discussion topic, and the usefulness of questions for students to develop new arguments. The authors reported that system-generated questions could not be distinguished from human-generated questions in the context of two discussion topics (topic about nuclear energy and topic about low interest rate) while the difference between system-generated questions and human-generated questions was noticed in the context of one discussion topic (defla-tion in Europe and US).

This paper not only investigates the quality of system-generated questions, but also the hypothesis that people of different populations perceive the quality of questions differently. That is, questions that are understandable, relevant to a discussion topic, and useful for teachers might be perceived not understandable, irrelevant to a discus-sion topic, not useful by students. This hypothesis will be investigated based on not only human-generated questions, but also questions that are generated by the question generation system developed by Le et al. (2014) for the domain of argumentation.

The remainder of this paper is structured as follows. The next section will review sources of semantic web data sources that can be used to generate questions. Then, in the third section, the study for investigating the formulated hypothesis will be described. In the fourth section, I will discuss on the results of the study, and in the final section, the conclusions will be summarized.

2   A Review of Semantic Web and Linked Open Data Sources

In order to help students develop new arguments for the argumentation, asking ques-tions is one of the useful strategies. In order to create questions, semantic information which is related to a given topic is required. Different semantic sources (such as semantic web data and linked open data) can serve to create questions. Presently, many useful semantic web and linked open data sources have been developed by large communities (including non-experts and experts), e.g., Wiktionary,[2] Openthesaurus,[3]

---

[2] http://de.wiktionary.org/wiki/Wiktionary:Hauptseite
[3] https://www.openthesaurus.de/

and GermaNet[4] for the German language; WordNet[5] and Freebase[6] for the English language; BabelNet[7] and DBPedia[8] are multilingual databases; and more.

In the following, I will review only the sources of semantic web data and linked open data for the English language that are maintained continuously and have a considerable number of datasets. The review is followed by a thorough analysis with respect to their usefulness in the context of question generation for argumentation.

The purpose of YAGO is combining information from different Wikipedia databases in multiple languages. The YAGO knowledge base is automatically constructed from Wikipedia and consists of entities, facts, and relations. Each article in Wikipedia represents an entity in the knowledge base YAGO. Two entities can stand in a relation. For example, the fact **AlbertEinstein** *hasWonPrize* **NobelPrize** has the relation *hasWonPrize* that has entities **AlbertEinstein** and **NobelPrize.** For the purpose of generating questions for helping develop new arguments, such relations may be useful. For example, we can generate a question using the relation *hasWonPrize:* "Which prize did Albert Einstein win?" This question may stimulate students to think about Einstein's work achievements for that prizes were announced. The version YAGO2 has over 9.8 million entities and 447 million facts (Hoffart et al. 2013). The YAGO3 version has 77 English relations (Mahdisoltani et al. 2015).

WordNet (Miller 1995) also provides a source of semantic information which can be related to a discussion topic. WordNet is an online lexical reference system for English. Each noun, verb, or adjective represents a lexical concept. A concept is represented as a synonym set (called synset), i.e., the set of words that share the same meaning. Between two synsets, WordNet provides semantic relations (12 relations for nouns). The hyponym relation represents a concept specialization. For example, for the concept "energy", WordNet provides a list of direct hyponyms which are directly related to the concept being searched and represent specializations: "activation energy", "alternative energy", "atomic energy", "binding energy", "chemical energy", and more. In addition, synsets can contain sample sentences to provide sample sentences, which can be used for generating questions. For example, if we input the word "energy" into WordNet, an example sentence, e.g., "Energy can take a wide variety of forms" for this concept is available. Sample sentences provided by WordNet may also be exploited to create questions, e.g.: "Which forms can energy take?" One of the advantages of WordNet is that it provides accurate information (e.g., hyponyms) and grammatical correct sample sentences which may serve useful semantic information for generating questions.

BabelNet (Navigli and Ponzetto 2012) is a multilingual semantic network which is an integration of lexicographic and encyclopedic knowledge from WordNet and Wikipedia. In addition to the standard WordNet relations, BabelNet is enriched with "gloss" relations and unlabeled relations that are derived from internal links in the

---

[4] http://www.sfs.uni-tuebingen.de/GermaNet/

[5] http://wordnet.princeton.edu/

[6] https://www.freebase.com/

[7] http://babelnet.org/

[8] http://dbpedia.org

Wikipages. A gloss relation is established based on a gloss for a concept in WordNet. For example, the gloss of the first synset of "play" is "a dramatic work intended for performance by actors on a stage", and so the first sense of "play" is gloss-related with the first sense of "actor" and the third sense of stage (in WordNet, each lexical unit may have several senses). Since Wikipages typically contain hypertext linked to other Wikipages, thus, it refers to related concepts. For instance, "play" (with sense "theatre") has links to "literature", "playwright", etc. BabelNet exploits these links in order to extend the relations between concepts in its database. In the current version (Navigli and Ponzetto 2012), BabelNet has 51,087,221 relations for the English language and this number of relations is enormously higher than the number of relations provided by WordNet (364,522). However, with respect to using gloss-relations and wikipages-relations for generating questions, I do not see benefits because these relations do not provide a specific semantic relationship between two concepts in order to generate a meaningful question. BabelNet is more useful than WordNet with regard to generating questions for different languages, because BabelNet is a multi-lingual database, while WordNet just supports the English language.

3   Do Different Populations Perceive Questions Differently?

The title of this section is the research question of this paper. Either questions are developed by human experts or generated by a computer systems, it is interesting for the community of question generation to know whether people of different populations perceive them differently. This research question is important because it helps us understand more about how people perceive questions, and thus, consequently, question generators (human or system) might have to adapt questions to target persons. In order to investigate this research question, I use the question generation system that has been developed by (Le et al. 2014) with the intention to support the process of argumentation. In this paper, I briefly summarize the approach to generating questions using WordNet developed by Le and colleagues (2014). The authors exploited two types of semantic information for generating questions. First, questions are generated using key concepts in a discussion topic. For example, the following discussion topic can be given to students in a discussion session:

"The catastrophe at the Fukushima power plant in Japan has shocked the world. After this accident, the Japanese and German governments announced that they are going to stop producing nuclear energy. Should we stop producing nuclear energy and develop renewable energy instead?"

From the discussion topic, the system extracts nouns and noun phrases to serve as key concepts for generating questions: *catastrophe, Fukushima power plant, nuclear energy, renewable energy*. These nouns and noun phrases are filled in a set of pre-defined question templates, and as a result, a set of questions are generated.

Table 1 shows a part of the set of pre-defined question templates implemented in the question generation system, where the left column represents the type of questions which can be instantiated by filling in the place-holder <X> of the corresponding template (the right column). The authors applied the question taxonomy developed by Graesser et al. (1992) and developed fourteen question templates.

Table 1. Question templates for question generation.

Type	Question template
Definition	What is <X>?
	What do you have in mind when you think about <X>?
	What does <X> remind you of?
Feature/Property	What are the properties of <X>?
	What are the (opposite)-problems of <X>?
	What features does <X> have?

The second type of information is using hyponyms provided in WordNet for each concept (cf. Sect. 2). Placeholders in pre-defined question templates can be filled with appropriate hyponym values for generating questions. For example, the noun "energy" exists in the discussion topic, and after extracting this noun as a key concept, it can be used as input for WordNet that provides several hyponyms, including "activation energy". The following question templates (Table 2, 2nd column) can be used to generate questions of the question type "Definition".

Table 2. Question templates and generated questions of the type "Definition".

Type	Question template	Question
Definition	What is <X>?	What is activation energy?
	What do you have in mind when you think about <X>?	What do you have in mind when you think about activation energy?
	What does <X> remind you of?	What does activation energy remind you of?

In addition to using hyponyms for generating questions, (Le et al. 2014) proposed to use example sentences provided by WordNet for each concept to generate questions. For example, the following questions have been generated using the sample sentence that is provided in WordNet "*catalysts are said to reduce the energy of activation during the transition phase of a reaction*".

- *Are catalysts said to reduce the energy of activation during the transition phase of a reaction?*
- *When are catalysts said to reduce the energy of activation?*
- *What are catalysts said to reduce during the transition phase of a reaction?*
- *What are said to reduce the energy of activation during the transition phase of a reaction?*
- *What are catalysts said to reduce the energy of during the transition phase of a reaction?*

Since in a pre-study Le et al. (2014) found that most questions generated using sample sentences provided by WordNet do not represent meaningful question items, this type of semantic information (sample sentences) for generating questions is opted out of the study that will be described in the next section.

3.1   Study Design

The goal of this study is to investigate how people from different populations perceive questions that are generated by human experts and by a computer system. For this purpose, the author invited three groups of human raters to join the study. The first group included seven computer scientists who are professors or PhD students of Computer Science. The second group is represented by six senior researchers of the argumentation and the question generation communities. The third group included six student teachers who are studying Computer Science Education at the Humboldt Universität zu Berlin and all of them are native Germans. They can understand English properly, because all German high school students must study English as the first foreign language.

The study consists of two phases. First, eight experts from the research communities of argumentation and question/problem generation (six of them are in the second population of human raters) were invited to manually create questions. They got the following three discussion topics by emails and were asked to create questions which can be used to support students in developing arguments. Since the eight experts work in the USA, Europe and Asia, the discussion domains were chosen with international relevance and had been in the news recently. For this study, the domains of energy and economy were chosen. Each discussion topic consisted of two sentences and an initial discussion question. The intention of this kind of construction for discussion topics was that the discussion participants and the human experts should have enough "materials" for thinking about a specific problem. If a discussion topic was too short (e.g., only a sentence or a discussion question), this might make it difficult for discussion participants to initiate a discussion or for human experts to think of questions to be generated:

Topic 1: *The catastrophe at the Fukushima power plant in Japan has shocked the world. After this accident, the Japanese and German governments announced that they are going to stop producing nuclear energy. Should we stop producing nuclear energy and develop renewable energy instead?*

Topic 2: *Recently, although the International Monetary Fund announced that growth in most advanced and emerging economies was accelerating as expected. Nevertheless, deflation fears occur and increase in Europe and the US. Should we have fear of deflation?*

Topic 3: *"In recent years, the European Central Bank (ECB) responded to Europe's debt crisis by flooding banks with cheap money...ECB President has reduced the main interest rate to its lowest level in history, taking it from 0.5 to 0.25 percent".*[9] *How should we invest our money?*

From eight experts, 54 questions for Topic 1, 47 questions for Topic 2, and 40 questions for Topic 3 were received.

Then, the same discussion topics were input into the question generation system for argumentation. For each discussion topic, the system generated several hundred

---

[9] http://www.spiegel.de/international/europe/ecb-surprises-economists-by-dropping-keyinterest-rate-to-historic-low-a-932511.html.

questions (e.g., 844 questions for Topic 1), because from each discussion topic several key concepts were extracted, and each key concept was extended with a set of hyponyms queried from WordNet. For each key concept and each hyponym, fourteen questions have been generated based on fourteen pre-defined question templates. Since the set of generated questions was too big for expert evaluation, a small amount of automatic generated questions was selected randomly, so that the proportion between the automatic generated questions and the human generated questions was about 1:3. There were two reasons for this proportion. First, in case the proportion between automatically generated questions and human generated questions is too high, then it could influence the real "picture" of human generated questions. Second, a trade-off between having enough (both human-generated and system-generated) questions for evaluation and a moderate workload for human raters needs to be considered. The proportion of automatic generated questions and human generated questions is shown in Table 3.

Table 3. Number of questions generated by human experts and by the system.

	Topic 1 No. of questions	Topic 2 No. of questions	Topic 3 No. of questions
Human-generated	54	47	40
System-generated	16	15	13
Total	70	62	53

In the second phase of the study, the whole list of human-generated and system-generated questions was given to the first group (computer scientists) and the third group (student teachers). For each individual human rater of the second group, an individual list of mixed questions was created, i.e., the questions which have been generated by each senior researcher of the argumentation and the question generation communities were removed, because they would identify the questions created by themselves easily. Thus, the list of questions to be rated by the second group was shorter than the list of questions for the first and the third groups. All raters were asked to rate each question based on a scale between 1 (bad) and 3 (good) over three quality criteria: the understandability of a question, the relevance of a question to a given discussion topic, the usefulness of a question for students to develop new arguments.

3.2 Results: Quality Difference Between System-Generated and Human-Generated Questions

In order to determine the quality difference between system-generated and human-generated questions, two-sided t-tests are used and p-values are taken as indicators for statistical significance of the difference in quality between the two groups.

Table 4 shows the quality of human-generated questions and system-generated questions for Topic 1. From this table we can notice that for the group of student teachers, the human-generated questions are significant better than system-generated questions with respect to their understandability ($t = 6.52$, $p < 0.0001$), their relevance to the given discussion topic ($t = 9.87$, $p < 0.0001$), their usefulness for helping students develop new arguments (t = 10.16, p < 0.0001). From the perspective of computer scientists and researchers of the argumentation and question generation communities, system-generated questions deserved better ratings with respect to understandability: the difference between system-generated questions and human-generated questions was statistically not significant (Group 1: t = 0.67, p = 0.51; Group 2: t = 0.04, p = 0.9642). With respect to the relevance of questions, the second and the third groups agreed that human-generated questions are significantly better than system-generated questions (Group 2: t = 5.67, p < 0.0001; Group 3: t = 9.87, p < 0.0001). With respect to the usefulness of questions, the ratings of three groups indicated that human-generated questions are significantly better than system-generated questions (Group 1: t = 0.39, p = 0.0009; t = 5.91, p < 0.0001; Group 3: t = 10.16, p < 0.0001). That is, all three groups agreed that system-generated questions are not useful as human-generated questions for helping students develop new arguments. In overall, the first and the second groups rated system-generated questions higher than average (since the rating scale is from 1 to 3, the averaged value is 1.5) whereas the group of student teachers only rated the understandability of system-generated questions higher than average (m = 2.13, s.d. = 0.79).

Table 4. Quality of questions for Topic 1

	Understandability mean (s.d.)	Relevance mean (s.d.)	Usefulness mean (s.d.)
Group 1: Computer scientists			
System-GQ	2.19 (0.89)	1.96 (0.87)	1.69 (0.69)
Human-GQ	2.28 (0.80)	2.14 (0.86)	2.12 (0.87)
Difference	t = 0.67	t = 1.25	t = 0.39
Significance	p = 0.51 (not significant)	p = 0.21 (not significant)	p = 0.0009 (significant)
Group 2: Argumentation and question generation researchers			
System-GQ	2.53 (0.57)	1.81 (0.59)	1.5 (0.76)
Human-GQ	2.53 (0.62)	2.51 (0.60)	2.35 (0.68)
Difference	t = 0.04	t = 5.67	t = 5.91
Significance	p = 0.9682 (not significant)	p < 0.0001 (significant)	p < 0.0001 (significant)
Group 3: Student teachers			
System-GQ	2.13 (0.79)	1.47 (0.72)	1.44 (0.67)
Human-GQ	2.82 (0.43)	2.65 (0.55)	2.67 (0.58)
Difference	t = 6.52	t = 9.87	t = 10.16
Significance	p < 0.0001 (significant)	p < 0.0001 (significant)	p < 0.0001 (significant)

Table 5 shows the quality of questions that have been generated for Topic 2 by human experts and by the system. The first point we can learn from results shown in this table is that all groups of human raters rated human-generated questions significantly better than system-generated questions. This result for Topic 2 is consistent with the conclusion of the evaluation study conducted by Le and Pinkwart (2015). The second point is that system-generated questions have been rated over average (e.g., Group 1 rated 2.4 for understandability, Group 2 rated 1.6 for the relevance criterion, Group 3 rated 1.73 for the relevance criterion) by three groups except the rating for the usefulness criterion given by the group of student teachers (1.37). The similar picture for human-generated questions can also be identified: with respect understandability, three groups rated between 2.61 and 2.76; for the relevance criterion, the ratings are between 2.43 and 2.56; and for the usefulness criterion the ratings given each group is almost the same (between 2.34 and 2.39).

Table 5. Quality of questions for Topic 2

	Understandability mean (s.d.)	Relevance mean (s.d.)	Usefulness mean (s.d.)
Group 1: Computer scientists			
System-GQ	2.40 (0.77)	1.83 (0.7)	1.87 (0.82)
Human-GQ	2.76 (0.48)	2.43 (0.73)	2.37 (0.70)
Difference	t = 3.01	t = 3.93	t = 3.29
Significance	p = 0.0031 (significant)	p = 0.0001 (significant)	p = 0.0013 (significant)
Group 2: Argumentation and question generation researchers			
System-GQ	1.97 (0.89)	1.60 (0.77)	1.50 (0.68)
Human-GQ	2.61 (0.64)	2.56 (0.67)	2.39 (0.72)
Difference	t = 4.21	t = 6.46	t = 5.90
Significance	p = 0.0001 (significant)	p = 0.0001 (significant)	p = 0.0001 (significant)
Group 3: Student teachers			
System-GQ	2.13 (0.86)	1.73 (0.58)	1.37 (0.56)
Human-GQ	2.74 (0.60)	2.47 (0.63)	2.34 (0.71)
Difference	t = 4.33	t = 5.63	t = 6.85
Significance	p = 0.0001 (significant)	p = 0.0001 (significant)	p = 0.0001 (significant)

Table 6 shows the quality of generated questions for Topic 3. First, we can see that the given ratings are not consistent among three groups of human raters. The ratings of the group computer scientists for system-generated questions with respect to the understandability (t = 0.33, p = 0.7397), the relevance (t = 1.54, p = 0.1272), and the usefulness (t = 1.89, p = 0.0613) are not significantly different from the ratings for human-generated questions. That means that system-generated questions are of similar quality as human-generated questions. On the contrary, the group of argumentation and question generation researchers held the human-generated questions significantly better than system-generated questions with respect to the relevance (t = 3.34, p = 0.0012), the usefulness (t = 3.67, p = 0.0004). Similarly, the group of student teachers rated the human-generated questions significantly better than the system-generated questions with respect to the understandability (t = 2.65, p = 0.0093) and the relevance (t = 2.70, p = 0.0081).

Table 6. Quality of questions for Topic 3

	Understandability mean (s.d.)	Relevance mean (s.d.)	Usefulness mean (s.d.)
Group 1: Computer scientists			
System-GQ	2.21 (0.72)	1.92 (0.88)	1.71 (0.69)
Human-GQ	2.27 (0.80)	2.21 (0.78)	2.04 (0.76)
Difference	t = 0.33	t = 1.54	t = 1.89
Significance	p = 0.7397 (not significant)	p = 0.1272 (not significant)	p = 0.0613 (not significant)
Group 2: Argumentation and question generation researchers			
System-GQ	2.25 (0.85)	1.92 (0.88)	1.50 (0.83)
Human-GQ	2.50 (0.74)	2.50 (0.68)	2.24 (0.85)
Difference	t = 1.37	t = 3.34	t = 3.67
Significance	p = 0.1754 (not significant)	p = 0.0012 (significant)	p = 0.0004 (significant)
Group 3: Student teachers			
System-GQ	2.38 (0.88)	2.17 (0.76)	2.25 (0.74)
Human-GQ	2.76 (0.51)	2.59 (0.59)	2.47 (0.68)
Difference	t = 2.65	t = 2.70	t = 1.39
Significance	p = 0.0093 (significant)	p = 0.0081 (significant)	p = 0.1682 (not significant)

Second, it is surprising that the ratings given by the group of student teachers for system-generated questions (understandability: 2.38, relevance: 2.17, usefulness: 2.25) are higher than the ratings given by the group Computer Scientists (understandability: 2.21, relevance: 1.92, usefulness: 1.71) over three criteria, while for Topic 1 and Topic 2, the group of student teachers rated system-generated questions always lower than human-generated questions. Why the phenomenon of disagreement on ratings between the three groups appeared and why the group of student teachers rated system-generated questions for Topic 3 better than for other discussion topics, these need further investigation.

3.3   Difference of Perceiving the Quality of Questions Between Three Populations of Human Raters

In this section, I investigate whether the quality of (both system-generated and human-generated) questions is perceived differently between three populations of human raters (computer scientists, researchers in argumentation and question generation communities, and student teachers). For this purpose, ANOVA will be used to analyze the difference of the quality of questions over three groups. ANOVA variance analysis will be performed over three independent samples. Each sample represents the ratings collected from each group of human raters. The samples are independent because for the group of argumentation and question generation researchers their own questions have been removed from the set of mixed questions to be rated. They should not rate the questions that have been generated by themselves. The other groups (computer scientists and student teachers) were assigned with a complete set of mixed questions.

Table 7 shows the difference of quality of system-generated questions (3rd–4th rows) and human-generated questions (6th–7th rows) perceived by three groups of human raters. These questions have been generated for the discussion topic about stopping nuclear energy (Topic 1). With respect to the understandability and the usefulness, while the difference in quality of system-generated questions between three groups of human raters is statistically not significant (understandability: $p = 0.08$, usefulness: $p = 0.26$), the difference in quality of human-generated questions between three groups of human raters is statistically significant ($p < 0.0001$ over all three criteria). This indicates that the three groups of human raters perceived the quality of human-generated questions for Topic 1 differently. Similarly, three groups of human raters rated the relevance of system-generated questions significant differently.

Table 7. Difference between three groups of human raters with respect to questions for Topic 1

	Understandability mean (s.d.)	Relevance mean (s.d.)	Usefulness mean (s.d.)
System-generated questions			
Difference	F = 2.6	F = 4.05	F = 1.38
Significance	p = 0.0789 (not significant)	p = 0.0201 (significant)	p = 0.2559 (not significant)
Human-generated questions			
Difference	F = 21.97	F = 18.46	F = 18.17
Significance	p < 0.0001 (significant)	p < 0.0001 (significant)	p < 0.0001 (significant)

Table 8 shows results of ANOVA analysis over three groups of human raters for the quality of system-generated and human-generated questions. These questions have been generated for the discussion topic about fear of deflation in Europe and the USA (Topic 2). For system-generated questions, with respect to the understandability and the relevance, there is no difference in ratings between three groups (understandability $p = 0.1388$, relevance: $p = 0.4226$). This indicates that all three groups of human raters agreed on the high understandability of system-generated questions (m = 1.97–2.40, cf. Table 5) and the moderate relevance (m = 1.60–1.83, cf. Table 5). However, with respect to the usefulness, there is a significant difference between three groups of human raters ($p = 0.0187$). This indicates that different groups of human raters perceived the usefulness of system-generated questions differently.

For human-generated questions, the ratings of human raters are consistent over all three criteria, i.e., there is no significant difference in ratings among three groups of human raters (6th–7th rows of Table 8). That is, they agreed on the high understandability (m = 2.61–2.76), high relevance (m = 2.43–2.56), and high usefulness (m = 2.29–2.39) (cf. Table 5).

Table 9 shows the difference in ratings between three groups of human raters for system-generated and human-generated questions. These questions have been developed for the discussion topic about the low interest rate in Europe (Topic 3). The table shows that with respect to the understandability and the relevance, there was no significant difference in ratings for system-generated questions between three groups of human raters (Table 9, 4th row: understandability: $p = 0.7642$, relevance: $p = 0.5001$). This indicates that all human raters agreed on the high understandability (m = 2.21–2.38, cf. Table 6) and high relevance (m = 1.92–2.17, cf. Table 6) of system-generated questions. However, with respect to the usefulness of system-generated questions, the difference in ratings between three groups is significant. This indicates that different groups of human raters perceived the usefulness of system-generated questions differently. We notice that this is also the case for Topic 2.

Table 8. Difference between three groups of human raters with respect to questions for Topic 2

	Understandability mean (s.d.)	Relevance mean (s.d.)	Usefulness mean (s.d.)
System-generated questions			
Difference	F = 2.02	F = 0.87	F = 4.17
Significance	p = 0.1388 (not significant)	p = 0.4226 (not significant)	p = 0.0187 (significant)
Human-generated questions			
Difference	F = 1.7	F = 0.9	F = 0.11
Significance	p = 0.1847 (not significant)	p = 0.4078 (not significant)	p = 0.8959 (not significant)

Table 9. Difference between three groups of human raters with respect to questions for Topic 3

	Understandability mean (s.d.)	Relevance mean (s.d.)	Usefulness mean (s.d.)
System-generated questions			
Difference	F = 0.27	F = 0.7	F = 6.29
Significance	p = 0.7642 (not significant)	p = 0.5001 (not significant)	p = 0.0031 (significant)
Human-generated questions			
Difference	F = 9.7	F = 6.41	F = 6.25
Significance	p < 0.0001 (significant)	p = 0.0020 (significant)	p = 0.0023 (significant)

For human-generated questions, the difference in ratings is significantly different over all three criteria (cf. 6th–7th rows of Table 9). This indicates that three populations perceived the quality of human-generated questions, which have been developed for Topic 3, differently, although the understandability (m = 2.27–2.76, cf. Table 6), the relevance (m = 2.21–2.59, cf. Table 6), and the usefulness (m = 2.04–2.47, cf. Table 6) of those questions were rated highly. This phenomenon is similar to the human-generated questions for discussion Topic 1 (cf. Table 7).

In summary (cf. Table 10), three groups of human raters perceived human-generated questions differently in the context of Topics 1 and 3 over three criteria. System-generated questions, they were perceived differently by three groups of human raters with respect to specific criteria: the relevance of questions for Topic 1 and the usefulness of questions for Topics 2 and 3.

Table 10. Summary of differences between three groups of human raters

	Understandability	Relevance	Usefulness
Topic 1			
System-GQ	No	Yes	No
Human-GQ	Yes	Yes	Yes
Topic 2			
System-GQ	No	No	Yes
Human-GQ	No	No	No
Topic 3			
System-GQ	No	No	Yes
Human-GQ	Yes	Yes	Yes

4 Related Work and Discussion

Several applications of question generation for argumentation have been devised. Liu and colleagues (Liu et al. 2012) introduced a system (G-Asks) for improving students' writing skills (e.g., citing sources to support arguments, presenting the evidence in a persuasive manner). The approach implemented in this system consists of three stages. First, citations in an essay written by the student are extracted, parsed and simplified. Then, in the second stage, the citation category (opinion, result, aim of study, system, method, and application) is identified for each citation candidate. In the final stage, an appropriate question is generated using pre-defined question templates. Evaluation studies have shown that the system could generate questions as useful as human supervisors and significantly outperformed human peers and generic questions in most quality measures after filtering out questions with grammatical and semantic errors (Liu et al. 2012).

While the work of Liu and colleagues (Liu et al. 2012) focused on improving the writing skills of students, Adamson and colleagues (2013) proposed to generate discussion questions automatically in order to support instruction. Adamson and colleagues investigated three different approaches of selecting sentences from a summary text: the cosine similarity (Huang 2008), LSA content scores (Dumais 2004), and TF-IDF uniqueness (Wu et al. 2008). Selected representative sentences are transformed into discussion questions. In order to rank the generated questions on the basis of abstraction and ability to trigger discussion, the authors calculated the subjectivity score by averaging the subjectivity values of each word in the sentence using SentiWordNet (Baccianella et al. 2010). SentiWordNet is a database of word-senses that provides subjectivity scores assigned to each word. Discussion questions have been generated applying these three approaches and evaluated by asking four teachers for rating. With respect to stimulating discussion, the LSA and Cosine similarity approaches were significantly better than TF-IDF. In addition, the evaluation study showed that there were no significant distinctions between the approaches on the dimension of comprehensibility (i.e., a generated question is comprehensible) and important themes (i.e., a generated question touches upon important themes from the story).

Similarly to the work of Adam and colleagues (2013), the approach of generating questions I present in this paper aims at stimulating the process of developing new arguments for a discussion topic. The difference between this approach and the approaches of Liu et al. (2012) and Adamson et al. (2013) is that the question generation approach being presented in this paper deploys WordNet as a source of semantic information for generating questions.

From the results of the study in the previous section, two lessons have been learned. First, in the context of a specific topic (e.g., Topic 3), system-generated questions have similar quality as human-generated questions over all three criteria (understandability, relevance, and usefulness). Second, different groups of human raters perceive, especially, human-generated questions differently. System-generated questions have also been perceived differently, but with respect to their relevance or their usefulness.

These results are important for instructors and automatic question generators to adapt questions to individual target groups (e.g., primary school students, high school students, or university/college students). For instructors, if their questions are not understandable for students, they may explain or reformulate them in another way. For automatic question generators, it is difficult to reformulate questions. I propose two strategies for automatic question generators. First, we can define different sets of question templates for different levels of target groups. The formulation of these questions templates depends on the level of the target group. If the target group consists of intellectual university students, then question templates can be formulated in a scientific manner. Second, we can adopt different question taxonomies for different target groups. Most school teachers know the Bloom's taxonomy (Bloom 1956), that has six levels: (1) knowledge, (2) comprehension, (3) application, (4) analysis, (5) synthesis, (6) evaluation. Teachers usually apply this question taxonomy in schools (Arias de Sanchez 2013). Beside the Bloom's taxonomy, there are many other question taxonomies, e.g. PREG (Otero and Graesser 2001), Schreiber (1967), Pate and Bremer (1967), among which the taxonomy developed by Graesser et al. (1992) is widely used for tutoring. To my best knowledge, until now, there is no study that compares the applicability of different question taxonomies for different target groups. However, question taxonomies may be used to individualize questions for automatic question generator.

During the second phase of the evaluation study (cf. Sect. 3.1), some student teachers optionally informed me about the different criteria they used to identify system-generated questions. Their criteria are very various: (1) a system-generated question is superficial with regard to a given discussion topic, (2) a system-generated question is similar to another one in the mixed set of questions, (3) a system-generated question that expects a factual answer and is intuitive (e.g., "What features does ECB president have?"), (4) a system-generated question that contains unknown information (e.g., "How will those policies affect those outcomes/stakeholders?"). To identify human-generated questions, they applied the criterion: human-generated questions may have typo/syntax errors, while system-generated questions are error-free. However, some of these criteria for guessing system-generated may be not justified, because for instance the developer of question templates could also have made typo/syntax errors as well.

Some of the human experts in the argumentation and question generation communities had more systematic criteria when they were asked to generate questions. Since I did not request all human experts to explicitly explain how they generated questions, I did not receive their strategies of generating questions. However, some of them optionally described their strategies. One of them applied different types of arguments/argumentative schemes based on the argumentum model of Rigotti and Greco Morasso (2010). In the following, I list all the argument types that have been used to generate questions to stimulate the argumentative reflection of students by one of the human experts in the argumentation community (the questions in the brackets were generated by her):

- Arguments of consequence or warning: e.g., "Think about the consequences of the nuclear catastrophe: which were (and still are) the consequences on people, in particular on their health condition?"
- Arguments of alternatives: e.g., "Think about different sources of renewable energy: which alternative sources of energy do we have/know?"
- Arguments of likeliness or difference: e.g., "Try to compare nuclear energy with sources of renewable energy: which are the differences in terms of productivity?"
- Arguments of termination and setting up: e.g., "Ponder about the possibility to go on producing nuclear energy: is it possible to make nuclear energy production safer than it is now?"
- Arguments of definition and ontological implications: "What does "deflation" mean? Is deflation bad for the economy of a country?"
- Argument of expert opinion/of authority: "Think about the authoritativeness of the information source: what is the International Monetary Fund? What does it do?"
- Argument of analogy: "Refer to similar past events: what did happen to the economies which had to face deflation?"
- Argument of final cause: "Think about the expected results of a money investment: what does an investor expect from investing money?"

Another human expert in the argumentation community explained how he generated questions. He applied several general questions when asking about a policy topic: "What stakeholders will be affected by this policy? What outcomes are the policy makers attempting to address? What other outcomes might be affected? How will this policy affect those outcomes/stakeholders? What is the evidence that this policy would affect these outcomes? What other policies are available (possible)? What are the pros/cons of each policy? Which policies have the best set of tradeoffs? For which stakeholders?" For example, for Topic 1 (nuclear energy), he generated the following questions: "What stakeholders will affected by stopping this production? What outcomes are the policy makers attempting to address? What other outcomes might be affected? How will stopping production affect those outcomes/stakeholders? What other policies are available (possible)? What are the pros/cons of each policy? Which policies have the best set of tradeoffs? For which stakeholders?" We notice that the last three questions are not related to a specific topic (e.g., nuclear energy) and can be used for general policy topics.

The argumentum model and the general questions for polity topics that were applied by two researchers of the argumentation community may be a good basis to define question templates for question generation systems that aim at generating questions to support argumentation. Whether the argumentum model and the general questions can be applied for different topics and results in meaningful questions, this needs further investigation.

5　Conclusions

In this paper, I have investigated the hypothesis that human-generated questions and computer-generated questions are perceived differently by different populations. For this purpose, a study in which eight human experts in the argumentation and question generation communities have been invited to construct questions and a question generation system has been deployed to generate questions. In total, 141 human-generated and 44 system-generated questions have been mixed and rated by three groups of human raters: computer scientists, argumentation and question generation researchers, and student teachers. The study confirmed the hypothesis for human-generated questions over three quality criteria (understandability, relevance, and usefulness) that different populations perceive questions differently. For system-generated questions, the hypothesis could only be confirmed on the criteria of relevance and usefulness of questions. The results of this study are the contribution of this paper that propose researchers on question generation to adopt different strategies (e.g., different question taxonomies, different sets of question templates) for different target groups of questions.

Acknowledgements. The author would like to thank researchers of the argumentation community and the problem/question generation community (Prof. Kevin Ashley, Prof. Kazuhisa Seta, Prof. Tsukasa Hirashima, Prof. Matthew Easterday, Prof. Reuma De Groot, Prof. Fu-Yun Yu, Dr. Bruce McLaren, Dr. Silvia De Ascaniis) for generating questions, Computer Scientists (Prof. Ngoc-Thanh Nguyen, Prof. Viet-Tien Do, Dr. Thanh-Binh Nguyen, Zhilin Zheng, Madiah Ahmad, Sebastian Groß, Sven Strickroth), and student teachers at the Humboldt-Universität zu Berlin for their contribution in this evaluation study. Especially, the author would like to express his gratitude to Prof. Pinkwart for introducing experts of the argumentation community.

References

Alkinson, J.M., Drew, P.: Order in Court. Macmillan, London (1979)

Adamson, D., Bhartiya, D., Gujral, B., Kedia, R., Singh, A., Rosé, C.P.: Automatically generating discussion questions. In: Lane, H., Yacef, K., Mostow, J., Pavlik, P. (eds.) AIED 2013. LNCS, vol. 7926, pp. 81–90. Springer, Heidelberg (2013)

Arias de Sanchez, G.: The art of questioning: using bloom's taxonomy in the elementary school classroom. Teach. Innov. Proj. 3(1), Article 8 (2013)

Baccianella, S., Esuli, A., Sebastiani, F.: SentiWordNet 3.0: an enhanced lexical resource for sentiment analysis and opinion mining. In: Proceedings of the Seventh International Conference on Language Resources and Evaluation (2010)

Barlow, A., Cates, J.M.: The impact of problem posing on elementary teachers' beliefs about mathematics and mathematics teaching. Sch. Sci. Math. **106**(2), 64–73 (2006)

Bloom, B.S.: Taxonomy of Educational Objectives: Handbook 1: Cognitive Domain. Addison Wesley Publishing, Boston (1956)

Brown, G., Wragg, E.C.: Questioning. Routledge, London (1993)

Chafi, M.E., Elkhouzai, E.: Classroom interaction: investigating the forms and functions of teacher questions in Moroccan primary school. J. Innov. Appl. Stud. **6**(3), 352–361 (2014)

Clayman, S., Heritage, J.: The News Interview: Journalists and Public Figures on the Air. Cambridge University Press, New York (2002)

Dillon, J.T.: Questioning and Teaching. A Manual of Practice. Croom Helm, London (1988)

Drew, P., Heritage, J.: Analyzing talk at work: an introduction. In: Drew, P., Heritage, J. (eds.) Talk at Work: Interaction in Institutional Settings, pp. 3–65. Cambridge University Press, New York (1992)

Dumais, S.T.: Latent semantic analysis. Annu. Rev. Inf. Sci. Technol. **38**(1), 188–230 (2004)

Graesser, A.C., et al.: Mechanisms that generate questions. In: Lauer, T., et al. (eds.) Questions and Information Systems. Erlbaum, Hillsdale (1992)

Heilman, M., Smith, N.A.: Question generation via over-generating transformations and ranking. Report CMU-LTI-09-013, Language Technologies Institute, School of Computer Science, Carnegie Mellon University (2009)

Hoffart, J., Suchanek, F., Berberich, K., Weikum, G.: YAGO2: a spatially and temporally enhanced knowledge base from wikipedia. Spec. Issue Artif. Intell. J. **194**, 28–61 (2013)

Huang, A.: Similarity measures for text document clustering. In: Proceedings of the 6th New Zealand Computer Science Research Student Conference, pp. 49–56 (2008)

Jouault, C., Seta, K.: Content-dependent question generation for history learning in semantic open learning space. In: Trausan-Matu, S., Boyer, K.E., Crosby, M., Panourgia, K. (eds.) ITS 2014. LNCS, vol. 8474, pp. 300–305. Springer, Heidelberg (2014)

Kunichika, H., Katayama, T., Hirashima, T., Takeuchi, A.: Automated question generation methods for intelligent english learning systems and its evaluation. In: Proceedings of the International Conference on Computers in Education, pp. 1117–1124 (2001)

Le, N.T., Nguyen, N.P., Seta, K., Pinkwart, N.: Automatic question generation for supporting argumentation. Vietnam J. Comput. Sci. **1**(2), 117–127 (2014). Springer Verlag

Le, N.T., Pinkwart, N.: Evaluation of a question generation approach using open linked data for supporting argumentation. Special Issue on Modeling, Management and Generation of Problems/Questions in Technology-Enhanced Learning, J. Res. Pract. Technol. Enhanc. Learn. (RPTEL) (2015)

Lin, L., Atkinson, R.K., Savenye, W.C., Nelson, B.C.: Effects of visual cues and self-explanation prompts: empirical evidence in a multimedia environment. Interact. Learn. Environ. J. **49**(1), 83–110 (2014)

Liu, M., Calvo, R.A., Rus, V.: G-Asks: an intelligent automatic question generation system for academic writing support. Dialogue Discourse **3**(2), 101–124 (2012)

Mahdisoltani, F., Biega, J., Suchanek, F.M.: YAGO3: a knowledge base from multilingual wikipedias. In: Proceedings of the Conference on Innovative Data Systems Research (CIDR 2015) (2015)

Miller, G.A.: WordNet: a lexical database for english. Commun. ACM **38**(11), 39–41 (1995)

Morgan, N., Saxton, J.: Asking better questions. Pembroke Publishers, Makhma (2006)

Mostow, J., Chen, W.: Generating instruction automatically for the reading strategy of self-questioning. In: Proceeding of the Conference on AI in Education, pp. 465–472 (2009)

Mostow, J., Beck, J.E.: When the rubber meets the road: lessons from the in-school adventures of an automated reading tutor that listens. In: Schneider, B., McDonald, S.-K. (eds.) Scale-Up in Education, pp. 183–200. Rowman and Littlefield Publishers, Lanham (2007)

Navigli, R., Ponzetto, S.P.: BabelNet: the automatic construction, evaluation and application of a wide-coverage multilingual semantic network. J. Artif. Intell. **193**, 217–250 (2012)

Otero, J., Graesser, A.C.: PREG: elements of a model of question asking. J. Cogn. Instr. **19**(2), 143–175 (2001)

Pate, R.T., Bremer, N.H.: Guiding learning through skillful questioning. Elem. Sch. J. **67**, 417–422 (1967)

Rigotti, E., Greco Morasso, S.: Comparing the argumentum model of topics to other contemporary approaches to argument schemes: the procedural and material components. Argumentation **24**(4), 489–512 (2010)

Rothstein, D., Santana, L.: Teaching students to ask their own questions. Harv. Educ. Lett. **27**(5) (2014). http://hepg.org/hel-home/issues/27_5/helarticle/teaching-students-to-ask-their-own-questions_507

Schreiber, J.E.: Teacher's question-asking techniques in social studies. Doctoral dissertation, University of Iowa, No. 67-9099 (1967)

Suchanek, F.M., Kasneci, G., Weikum, G.: YAGO: a core of semantic knowledge unifying WordNet and wikipedia. In: Proceedings of the International World Wide Web Conference, pp. 697–706, ACM (2007)

Tenenberg, J., Murphy, L.: Knowing what i know: an investigation of undergraduate knowledge and self-knowledge of data structures. Comput. Sci. Educ. **15**(4), 297–315 (2005)

Wu, H.C., Luk, R.W.P., Wong, K.F., Kwok, K.L.: Interpreting TF-IDF term weights as making relevance decisions. ACM Trans. Inf. Syst. **26**(3), 13:1–13:37 (2008)

Yu, F.Y., Pan, K.J.: The effects of student question-generation with online prompts on learning. Educ. Technol. Soc. **17**(3), 267–279 (2014)

Reflection of Intelligent E-Learning/Tutoring - The Flexible Learning Model in LMS Blackboard

Ivana Simonova, Petra Poulova[(✉)], and Pavel Kriz

University of Hradec Králové, Rokitanského 62,
Hradec Králové, Czech Republic
{ivana.simonova,petra.poulova,pavel.kriz}@uhk.cz

Abstract. The article encompasses the theoretical background and practical concept of teaching/learning through online courses as an example of smart solution of e-learning system adjusting to individual learning preferences, including students' reflection on the individualized online instruction. First, the 'Unlocking the Will to Learn' concept by C.A. Johnston is introduced and implemented in the research design. Second, the pedagogical experiment reflecting individual learning style preferences is conducted. For this phase of research the e-application was designed which considers learners' individual characteristics and consequently generates the online course content adjusted to them. Finally, learners' feedback after studying in the course is presented.

Keywords: Cognitive modelling · eLearning · Intelligent e-learning/tutoring · Tracking · Ubiquitous environment

1 Introduction

The process of instruction, especially the technology enhanced learning, is highly appreciated by most learners of all age-groups. Fast technical development and new technologies, globalization of the world, the need of unlimited access to education for everybody – these are some of the reasons which enabled and caused information and communication technologies (ICT) were brought to all spheres of everyday life, including the field of education.

Adult students often accept this approach as the only way how to study and work, and reach new competences. The younger the learners are, the more easily they accept ICT in everyday life, including education. Being called 'digital natives' by Prensky [1], they differ substantially from previous generations of 'digital' immigrants. They have not changed their behaviour (slang, clothes, body adornments etc.) only, as it happened between generations before, but a really big discontinuity, singularity was detected with them, caused by the appearance and rapid dissemination of digital technology at the end of the 20th century. As a result of this ubiquitous environment and the sheer volume of their interaction with it, current learners think and process information fundamentally differently from their predecessors and differences go far further and deeper than most educators suspect or realize, Berry states (in [2]): *"Different kinds of*

© Springer-Verlag Berlin Heidelberg 2015
N.T. Nguyen (Ed.): Transactions on CCI XVIII, LNCS 9240, pp. 20–43, 2015.
DOI: 10.1007/978-3-662-48145-5_2

experiences lead to different brain structures. ... Today's students' brains have physically changed as a result of the environment they grew up, their thinking patterns have changed".

2 Research Design

This fact provides impact not only on a single learner but the whole education systems are affected, and curricula should reflect this state, particularly in the field of teaching methods where the above mentioned ICT can help substantially.

The question is whether tailoring the process of instruction running within the LMS to student's individual learning preferences can result in increasing the knowledge. To discover this was the main objective of the three-year research on 'A flexible model of the technology enhanced educational process reflecting individual learning styles' which was solved at the Faculty of Informatics and Management, University of Hradec Kralove, Czech Republic.

2.1 Theoretical Background

Within the last two decades the society changed substantially. The information and communication technologies have been implemented in all spheres. The development towards democracy and information and knowledge society transformed the existing structure of the Czech educational system, defined new competences reflected in the learning content, called for new teaching and learning strategies.

The effectiveness of the educational process, been given by such factors as learner's intelligence, prior knowledge, level of motivation, stress, self-confidence, and learner's cognitive and learning style, has been researched by numerous scientists from many countries. It is generally acknowledged that the instructor's teaching style should match the students' learning styles. Felder says that mismatching can cause a wide range of further educational problems [3]. It favours certain students and discriminates others, especially if the mismatches are extreme. On the other hand, if the same teaching style is used repeatedly, students become bored. Gregorc claims that only individuals with very strong preferences for one learning style do not study effectively, the others may be encouraged to develop new learning strategies [4]. Only limited number of studies have demonstrated that students learn more effectively if their learning style is accommodated. Mitchell concludes that making the educational process too specific to one user may restrict the others [5]. The possibility of individualization the educational process from the both students' and teachers' point of view is its greatest advantage [6].

Numerous scientists and researchers have been dealing with the role and impact of cognitive and/or learning styles within the process of instruction, but there are a fewer ones focusing on this field under the conditions of ICT-supported learning. More than a decade ago, Honey was one of those who were asked their opinion on learning styles in e-learning, specifically *'Are there e-learning styles?'* [7]. In 2000 he ran a research to investigate the existence or non-existence of e-learning styles. In the sample group of 242 respondents he indicated their individual learning style preferences and reacted to a long list of potential likes and dislikes about e-learning. Unfortunately, the correlation to the learning preferences did not reveal the significant differences as Honey expected - the likes and dislikes were remarkably similar regardless of learning style preferences [8]. Drilling down into deeper analysis another question appeared, i.e. whether people with different learning style preferences had the same things in mind when they signed up for these likes and dislikes. It seems unlikely that learning 'at my own pace' and 'when and for how long' would be the same for learners with different learning preferences. Honey concluded that despite his initial survey had failed to reveal e-learning styles as such, it discovered some important differences about how people approach online learning. *'One size fits all' has never worked for clothes. Why should it for e-learning?'* [7]. On the contrary, Honey criticized e-learning industry's developers and vendors for tending to be didactic rather than involving, preferring to tell rather than ask [9].

2.2   Individual Learning Preferences

Numerous researches have been conducted on how individual learning preferences to be efficiently accommodated. The Learning Combination Inventory (LCI) designed by C.A. Johnston was applied in this research to detect the learning style pattern of each learner [10]. To describe the whole process of learning, Johnston uses the metaphor of a combination lock saying that cognition (processing), conation (performing) and affectation (developing) work as interlocking tumblers; when aligned they unlock individual's understanding of his/her learning combination. The greater the intelligence, the more a child can learn. She attracts attention to the verb can, as no one says will learn. The LCI consists of 28 statements, responses to which are defined on the five-level Likert scale, and three open-answer questions:

- What makes learning frustrating for you?
- How would you like to show the teacher what you know?
- How would you teach students to learn?

The results are provided in the form of the four-figure score to quantitatively and qualitatively capture a student's cognitive, conative, and affective interactive learning combination (pattern). In other words, the score represents the degree to which each learner 'Uses first', 'Uses as needed', or 'Avoids' certain type of learning material and learning strategy, i.e. what his/her learning preferences are. Learners' responses create the schema (pattern) that drives the will to learn. The patterns are categorized into four groups and described as follows [10]:

- *sequential processors*, defined as the seekers of clear directions, practiced planners, thoroughly neat workers;
- *precise processors*, identified as the information specialists, info-details researches, answer specialists and report writers;
- *technical processors*, specified as the hands-on builders, independent private thinkers and reality seekers;

– *confluent processors*, described as those who march to a different drummer, creative imaginers and unique presenters.

The LCI differs from other widely used inventories (e.g. [11, 12]) as it focuses on how to unlock and what unlocks the learner's motivation and ability to learn, i.e. on the way how to achieve student's optimum intellectual development. This was the main reason why the LCI, not any traditional tool was applied for detecting respondents' individual learning styles within the above mentioned project.

2.3 Methodology of Research

The course of research was structured into following phases:

1. Pre-research activities, i.e.

   – Analyzing the "Learning Combination Inventory", translate and adapt it for the Czech educational system. Run a pilot research; consider the results.
   – Designing the online course and pilot it.
   – Preparing and piloting the questionnaire monitoring students' preferences in study materials (Q1), piloting the questionnaire, considering comments
   – Designing an e-application.
   – Designing and piloting the evaluation questionnaire (Q2).

2. Pedagogical experiment.

   – Setting the research sample, experimental and control groups.
   – Applying the LCI and determining students' learning styles.
   – Applying the questionnaire monitoring students' preferences in study materials (Q1), collecting and processing the data; interpreting the results.
   – Specifying the methodology, planning the process of instruction.
   – Designing and piloting didactic tests (pre-test, post-test).
   – Testing the entrance level of students' knowledge by a didactic test (pre-test).
   – Running the multistyle process of instruction.
   – Testing the students' level of knowledge after the instruction (post-test).
   – Comparing the results reached via different learning strategies.
   – Processing the collected data; interpreting the results.

3. Students' feedback after instruction

   – Applying the evaluation questionnaire (Q2), collecting and processing the data; interpreting the results.

   Several research methods were applied within the project:

– pedagogical experiment running on the "pre-test – instruction – post-test" structure,
– method of testing knowledge,
– method of questionnaire,
– statistic methods (t-test, Kolgomorov-Smirnov (K-S) test, Mann-Whitney test (Z-value).

To reach the research objectives following research tools were used:

- didactic tests to evaluate students' entrance level of knowledge before the process of instruction starts (pre-test) and students' final level of knowledge after the instruction (post-test);
- Learning Combination Inventory (LCI) to detect students' individual learning styles;
- questionnaire monitoring students' preferences in study materials (Q1);
- questionnaire monitoring students' feedback after the process of instruction (Q2);
- statistic software NCSS 2007 and MS Office Excel to process the collected data.

2.4   Process of Instruction

The process of instruction was held within the online course intentionally designed for this purpose in three versions:

- Version LCI reflecting the learner's style (experimental group 1) where students were offered such study materials, exercises, assignments, ways of communication and other activities which suit their individual learning styles; the selection was made electronically by an e-application which automatically generates the "offer"; this smart solution provides each student with types of materials appropriate to his/her learning style,
- Version CG providing all types of study materials to the learner, the process of selection is the matter of individual decision, the choices are monitored and compared to expected preferences defined by the LCI (experimental group 2).
- Version K reflecting the teacher's style (control group) where participants study under traditional conditions, when their course is designed according to the teacher's style of instruction which they are expected to accept.

The on-line course was designed in the LMS Blackboard. The content focused on library services, which is a topic students have to master before they start studying but they often have hardly any system of knowledge and skills in this field. The e-course was structured into eight parts covering the crucial content, i.e. Basic terminology, Library services, Bibliographic quotations, Electronic sources, Bibliographic search services, Writing professional texts, Writing bachelor and diploma theses and Publishing ethics.

2.5   Sample Group for Pedagogical Experiment

The sample group consisted of students of Faculty of Informatics and Management, University of Hradec Kralove, who enrolled in the online course running in the LMS Blackboard. It was designed as the learning environment so it provides all tools necessary for simulating the process of instruction efficiently.

All students were randomly (by lots) divided in three groups described below.

Nearly 400 respondents started the pedagogical experiment but only 324 finished it, from various reasons. Starting and final amount of respondents are presented in Table 1.

Table 1. Amounts of respondents

Group	Respondents before pedagogical experiment (n)	Respondents after pedagogical experiment (n)
K	130	113
CG	131	103
LCI	131	108
Total	392	**324**

3 E-Application Generating the Course Content

The application (plug-in) supporting the flexible model of instruction within the LMS was designed. Its main objective is to re-organize the introductory page of the e-course where the course content is presented to students. The criterion under which the application worked was the student's individual learning style. Categorization of learner's preferences (i.e. his/her individual learning style pattern) to certain types of learning style, which are mentioned above (sequential, precise, technical confluent processors), are not presented in the binary way (yes/no) but described by the fuzzy value expressing the relevance rate of each learner to a given group. Various types of learning materials are presented in such order which accommodates student's preferences, i.e. the plug-in arranges single items of the course content on the introductory page in such order which reflects the student's individual learning style pattern.

To design the online course reflecting learner's preferences as described above (LCI version), not only data on each student's learning style were required but also single items of the course content were classified according to the suitability (appropriateness) for a certain learning style, i.e. whether the material is preferred, accepted or refused by the student. Then, single types of study materials and activities were matched to each student's learning style pattern and the course was tailored to the individual student's needs. This final phase was carried out by the e-application (plug-in).

3.1 Process of Implementation

The plug-in is implemented as the extension of Building Block type for the BlackBoard Learn system. The administration rights are required for installing the extension in the system. The plug-in is distributed in the form of WAR file and script in JavaScript language. The WAR file having been installed by the administrator, the plug-in is available to course designers to be inserted the Course Content page by Add Interactive Tools. The plug-in creates a course item with static HTML code which contains:

- link to jQuery library hosted at http://ajax.googleapis.com;
- JavaScript code `jQuery.noConflict()` preventing from collision between the Prototype library, internally used by BlackBoard, and the `jQuery` library, used by the plug-in;
- link to `data/script.jsp` file which is part of the `Building Block` (plug-in);
- HTML DIV element where the new learning content reflecting learner's preferences is dynamically generated.

A new item `Table of Contents` is added to the main course menu. This item opens the entry page of the course where the plug-in is inserted, i.e. where study materials and learning activities are structured reflecting learner's preferences. Then, the original course content (folders with study materials and activities) is available under another item (Course Content) in the main course.

3.2   How the Plug-in Works

The plug-in is activated in learner's browser after accessing the `Table of Contents` page (where the plug-in is inserted). Then, the plug-in runs following activities:

- it downloads the jQuery library from the Internet (the `jQuery` library supports further activities);
- it downloads the JavaScript code generated by the `data/script.jsp` file;
  - this file generates `JSON` data providing information on the classification of files with learning objects, learner's preferences detected by LCI, evaluation of in/adequacy of learning objects to learner's preferences and adds the JavaScript code read from the `/uhk-flexible-learning/script.js` file within the given course;
- it activates the JavaScript code which calculates the in/adequacy rate of each learning object for each learner reflecting the LCI results;
- generates the `Table of Contents` and displays it in the place of DIV element, i.e. single learning objects (study materials and related activities) are presented in such order which reflects learner's LCI results, i.e. individual preferences.

If in case of error the `Content page` is not generated, the error notice appears.

3.3   Requirements for Plug-in Work

Under the `Control Panel – Content Collection`, resp. Files in the latest version, the `uhk-flexible-learning` folder should be created and the script.js and students.csv files uploaded.

Single topics in the `Table of Contents` are structured into folders, one topic per folder, and the link to each learning object (study material, activity) is included.

Each learning object in the folder is described by four figures of the value of -1, 0, 1 which correspond to four types of processors by Johnston's concept (sequential, precise, technical and confluent) as follows:

- minus one (−1) means this type of study material, activity, assignment, communication etc. is refused, i.e. does not match the given learning style;
- zero (0) is the middle value, i.e. the student neither prefers, nor refuses, but accepts this type;
- one (1) means this type is preferred, it matches the given learning style.

This three-state model could be extended to a wider scale of fuzzy values reflecting the Johnston's model in deeper detail. The above mentioned file students.csv contains the classification of students' individual learning patterns (i.e. fuzzy values reflecting the relevance rate of each learner pattern to a given group of processors – sequential, precise, technical confluent) as displayed in Table 2.

Table 2. Classification of students' individual learning patterns.

User name	Classification 1	Classification 2	Classification 3	Classification 4
krizpal	25	18	14	20
webct_demo_69259477001	20	12	18	27

Data are taken from the spreadsheet (e.g. MS Office Excel) in the CSV format, separated by semicolon, e.g. `krizpal;25;18;14;20`.

For designer view the designer user name is included in the `students.csv` file. If missing, the error notice is displayed saying the `Table of Contents` does not reflect student's preferences.

The plug-in requires student access to the Internet so that the `jQuery` library could be downloaded from the ajax.googleapis.com server.

The plug-in was designed for and tested in the Blackboard Learn system, version 9.1 and uses the Application Programming Interface (API) for detecting the Course Content, metadata for learning object classification, student's user name and for reading the `script.js` and `students.csv` files. Some functions of the API are not documented (e.g. reading metadata) and changes in Blackboard version are expected to require modification of Java code using the API.

3.4 Implementation Details

The process of plug-in implementation in the course is structured in several steps represented by activities of single files.

Step 1 is created by `create.jsp`, `create_proc.jsp` and `modify.jsp` files. This step is applied only once, in the moment of plug-in implementation in the course. The `create.jsp` and `create_proc.jsp` files contain codes in Java language which work for creating the above described item `Table of Contents`. Following classes of BlackBoard API are used:

- `Content` – the learning object in the BlackBoard system; the object is always assigned to a given course and it may belong to the folder;
- `FormattedText` – serves for creating the Table of Contents formatted by HTML code;

- `ContentDbPersister` – serves for saving the learning objects.

The `modify.jsp` file displays information the plug-in generated content (`Table of Contents`) cannot be adjusted but deleted if needed.

Step 2 is introduced by `script.jsp` file containing the code in Java language generating the necessary code in JavaScript which is subsequently interpreted by user browser. The `script.jsp` file processes the original `Course Content` and information on the user and generates data in the `JSON` format. The `script.js` file is then appended to the generated `JSON` data to be submitted together to the user's browser.

Compared to the previous version, which was designed for previous generations of the LMS WebCT/Blackboard, the current version is better implemented regarding to API, which was missing in WebCT. Thus the main problem of version 1 was eliminated, i.e. the plug-in dependence on concrete structure of HTML pages, as the Learning Content was detected by parsing HTML pages.

Step 3 works with `script.js` file which contains the main JavaScript code. The `script.js` is a common file saved in the course which can be easily adjusted by course designer so that higher flexibility was reached – if small changes in plug-in functionality are required, adjustments in script.js file are made and the plug-in re-installation is not necessary. The plug-in can be also tailored to the course requirements as each course has its own the `script.js` file.

The key part of the `script.js` file is the algorithm calculating the appropriateness of a learning object for the given student which is based on both the learning object and student classification (LCI pattern). The core of algorithm in JavaScript language is described below:

```
var totalEval = 0;
for (var i = 0; i < topicData.classification.length; i++)
{
    // rejected
    if (userData[i + 1] < refuseValue)
        totalEval += topicData.classification[i] *
                     (userData[i + 1] - refuseValue);
    // appreciated
    if (userData[i + 1] > acceptValue)
        totalEval += topicData.classification[i] *
                     (userData[i + 1] - acceptValue);
}
```

The algorithm in the cycle goes through single values of the given learning object classification in array `topicData.classification` (indexed from 0) and reflecting the `userData` (indexed from 1) it detects for each value whether the student refuses, accepts or prefers material of this type. The threshold values for decision-making process of accepting/refusing the type are saved in constants `refuseValue` and `acceptValue`. The appropriateness value of the type for the learner's LCI pattern is added to `totalEval` value (being 0 at the beginning). The final appropriateness rate is expressed by the `totalEval` variable. Then, the script.js file ranks learning objects in each folder according to the calculated rate and displays them to the student – the preferred types of learning materials and activities are on the top of the list, underlined, written in bold font of large size and black colour. A sample of individually re-organized table of contents is displayed in Fig. 1.

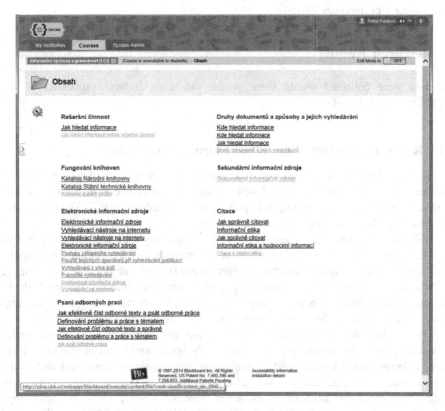

Fig. 1. The individually re-organized table of contents

The `script.js` file uses the `jQuery` library version 1.4.2 mainly for manipulating with the page content. In `jQuery` library the "$" function cannot be used as it colligates with the same one in the Prototype library used by Blackboard. That is why the `jQuery` function instead of $ function is used in the `script.js` code.

3.5   Currently Known Limits and Future Work

While designing the plug-in, several limits have been discovered for the time being. We have not succeed in hiding the item with original course content (which does not reflect individual learning preferences) in the main menu therefore we have renamed it

Course Content (as mentioned in chapter Process of implementation) and shifted it on the bottom position so that students did not primarily use it.

Currently the plug-in supports two-level hierarchy of learning objects, i.e. single learning materials and activities are presented in the form of files which are clustered into folders, one folder per topic. The more-level hierarchy requires changes in data structure in JSON format (in the Table of Contents) generated by the data/script.jsp file and in algorithms creating the new learning content reflecting learner's preferences (Table of Contents).

4 Research Results

All data were collected within three phases the research were processed by the NCSS2007 statistic software.

4.1 Results in Detection of Individual Learning Preferences

Before the pedagogical experiment started, individual learning style of each respondent had been detected by the Learning Combination Inventory. The structure of research groups from the point of learning style pattern is displayed in Table 2. The groups did not differ significantly, they were considered equal.

Following the research design, individual learning style of each respondent had been detected by the Learning Combination Inventory before the pedagogical experiment started. The LCI results structure of each group is displayed in Table 3 and Fig. 2.

Within the online course various types of learning materials were available to learners. The correlations between single learning style patterns and types of learning materials used in the online course were detected. Results are presented in Table 4.

The results (the recommended value of the correlation is 0.15 min.) show, *sequential* processors mostly use electronic study texts, books and professional literature, video-recorded lectures and presentations; they reject self-tests and other supportive materials, e.g. dictionaries. *Precise* processors work with books and professional literature, animations, examples, electronic study texts and other supportive materials, e.g. dictionary; they do not like video-recorded lectures. *Technical* processors often use animations and video-recorded lectures; they do not work with electronic study texts, other supportive materials, e.g. dictionaries and presentations. *Confluent* processors work with books and professional literature and self-test; they do not use electronic study texts, video-recorded lectures, presentations and other supportive materials, e.g. dictionaries.

4.2 Results of Pedagogical Experiment

The individual learning patterns in three research groups having been detected, the pedagogical experiment started following the "pre-test – instruction – post-test structure". The data collected from didactic tests were applied in the process of verifying hypotheses.

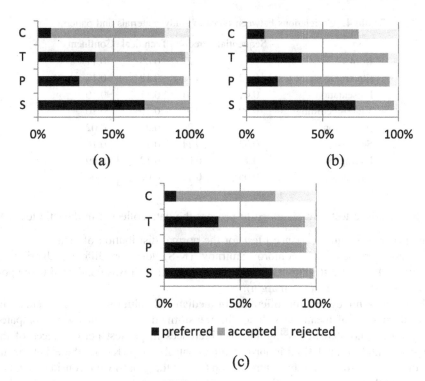

Fig. 2. Learning styles structure in the LCI (a), CG (b) and K (c) Groups

Table 3. Learning styles structure in LCI, CG and K groups

Group/n		Sequential	Precise	Technical	Confluent
LCI 108	Preferred	75	29	40	9
	Accepted	33	75	63	82
	Rejected	0	4	3	17
CG 103	Preferred	76	23	39	14
	Accepted	26	76	59	64
	Rejected	3	6	7	27
K 113	Preferred	80	26	41	9
	Accepted	31	80	64	74
	Rejected	2	7	8	30

Two hypotheses to be verified were defined as follows:

H_1: Students reach higher increase in knowledge if the process of instruction is adjusted to their learning style (LCI group) in comparison to the process reflecting teacher's style of instruction (group K)

H_2: Students reach higher increase in knowledge if they can study independently using all types of provided study materials (CG group) in comparison to the process reflecting teacher's style of instruction (K group)

Table 4. Correlations between types of study materials and patterns

	Sequential	Precise	Technical	Confluent
Books	0.11	0.27	0.05	0.11
Electronic study text	0.12	0.11	−0.18	−0.17
Presentation	0.01	0.01	0.11	−0.10
Video-recording	0.09	−0.03	0.20	−0.16
Animation	0.01	0.24	0.23	−0.02
Self-test	−0.04	0.11	0.12	−0.14
Examples	−0.11	0.00	−0.12	0.09
Dictionary	0.05	0.12	0.02	−0.04

Three statistic tests were applied to process the data collected in didactic tests:

- the parametric equal variance t-test for the normal distribution of data;
- the non-parametric Kolgomorov-Smirnov (K-S) test for different distribution (despite the total distribution of datawas normal, this test was conducted to support those situated close to the margin);
- the Mann-Whitney test for difference in medians (Z-value) was applied. Then, the transformation of means was done, the transformed mean values were compared and proved no statistically significant differences in pre-test performances of the experimental group 1 (LCI group) and the control group (K) and the experimental group 2 (CG group) and the control group (K), so the groups were considered equal. This entitled us to running the pedagogical experiment. Results are displayed in Table 5.

Table 5. Results of pre-tests

	CG	K	LCI
Mean	22.61	22.48	22.46
Min	6	13	6
Max	28	28	28
Range	22	15	22
SD	3.62	3.73	3.98
Modus	24	23	–
Median	24	23	23
t-test	−0.2506 (crit. 1.9706) = NR		–
	–	0.0366 (crit. 1.9704) = NR	
K-S test	0.16648 (crit. 0.086) = NR	0.1662979 (crit. 0.08) = NR	0.1451381 (crit. 0.084) = NR
Z-value	0.3717 = NR		–
	–	0.1826 = NR	

Note: NR: H_0 not rejected

After pre-testing the instruction started in three versions (groups) of the online course Library services – Information competence and education with the face-to-face entrance tutorial and closed with another one where learners' knowledge was tested. The period of study was three weeks of independent study. Students were randomly divided into three groups (LCI, CG, K) which differed in the extent of individualization of the process of instruction. After that, students' final knowledge was tested by post-test. The results are displayed in Table 6.

Table 6. Results of post-tests

	CG	K	LCI
Mean	26.34	25.42	26.10
Min	14	12	14
Max	30	30	30
Range	16	18	16
SD	2.98	4.13	2.42
Modus	28	28	28
Median	27	27	27
t-test	-1.8953 (crit. 1.9706) = NR	$-$	
	$-$	-1.4987 (crit. 1.9704) = NR	
K-S test	0.1875374 (crit.0.086) = NR	0.1783263 (crit. 0.08) = NR	0.1622858 (crit. 0.084) = NR
Z-value	1.5995 = NR		$-$
	$-$	0.1863 = NR	

Note: NR: H_0 not rejected

The collected data underwent the same procedure as the pre-tests did, i.e. the t-test, Kolgomorov-Smirnov and Mann-Whitney tests were used, so that test scores were compared: the LCI group (experimental group 1) to the control group (K) and the CG group (experimental group 2) to the control group (K). Results are displayed in Fig. 3 (K – LCI pre-test upper left; K – LCI post-test upper right; K – CG pre-test bottom left; K – CG post-test bottom right).

The results proved differences neither in pre-tests results, nor in post-tests performance. Thus it can be concluded *no statistically significant differences were discovered in students' knowledge either their learning preferences are reflected within the process of instruction (LCI version), whether they work independently being provided all types of study materials and activities (CG version) or the process follows teacher's style of instruction (K version).*

4.3   Results in Respondents' Evaluation of Online Study

The pedagogical experiment having been closed, the final evaluation questionnaire (Q2) was applied to collect data on students' opinions and experience from online learning in the course they attended. The questionnaire consisted of 22 items which monitored both the learners' didactic experience and technological problems the

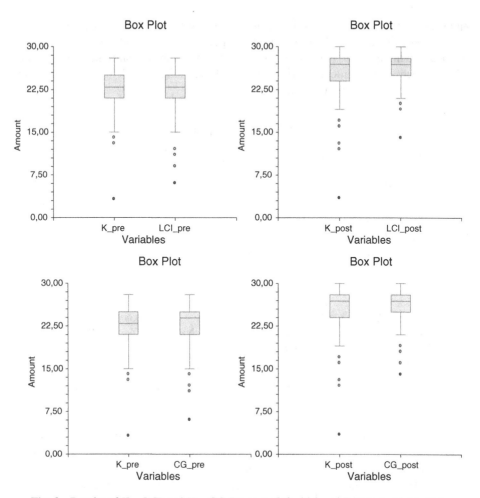

Fig. 3. Results of K – LCI and K – CG Pre-tests (left side) and Post-tests (right side)

learners might have had within the process of autonomous online learning. The detailed information about the sample group was collected in five items and 17 questions covered respondents' experience in studying the online course; seven statements were evaluated on the four-level Likert scale, six ones were the open-answer questions and four items were of multiple-choice type.

In the first part, the research group of 324 respondents was described in detail:

– 60–63 % of male respondents in each group (LCI, CG, K);
– respondents were from 20–50 years old, 80 % in the 20–24 year-old group;
– 62–67 % of respondents graduated from secondary professional schools, the others were grammar school graduates (29–45 %);
– 60–65 % of respondents did not have any experience in studying online courses, approximately 20 % of them had studied one course and 5 % were experienced online learners having passed four or more courses.

In the second part of the questionnaire the items focused mainly on evaluating the process of instruction considering e.g. the difficulty level of single topics, learners' feelings within the course of study etc. The difficulty was evaluated by on the seven-level scale from value 1 – least difficult topic (the structured pattern), value 2 – light grey colour) to value 7 – most difficult topic (black colour).

Creating quotations and Professional writing were considered the most difficult ones (Fig. 4).

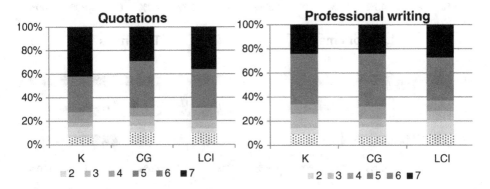

Fig. 4. Difficulty of topics (1 – least difficult, 7 – most difficult)

Creating quotations was considered the most difficult topic (level 7) by 40 % of respondents in the control group (K) reflecting the teacher's style of instruction and 25 % on level 6), followed by the LCI group (experimental group 1) in which individual learning styles were reflected in the process of instruction (35 %, 25 %) and by the CG group (experimental group 2; 29 %, 16 %).

Professional writing was also recognized a difficult topic; 26 % and 28 % of respondents in the control group evaluate it very difficult (levels 7 and 6), and similar results appeared in other groups: respondents in the LCI group showed 27 % and 25 %, in the CG group the results were 24 % and 31 %. The complete data are presented in Fig. 4.

The course of study was also evaluated from the point of learners' problems, difficulties and limits. Five criteria were set as follows:

- to start studying (i.e. find motivation, time etc.),
- to keep studying (i.e. keep motivated, have time etc.),
- lack of time (within the process of study),
- tiredness (within the process of study),
- problems with technology (tools not working properly etc.).

Data were evaluated on the six-level scale from no problems (level 1) to crucial problems (level 6). Results are presented in Fig. 5.

The results show that half of respondents (48 %) had no or little problems (levels 1–3) to make efforts and start studying in the LCI group while slightly fewer ones were detected in CG (46 %) and K (44 %) groups; higher rate was even expressed under the

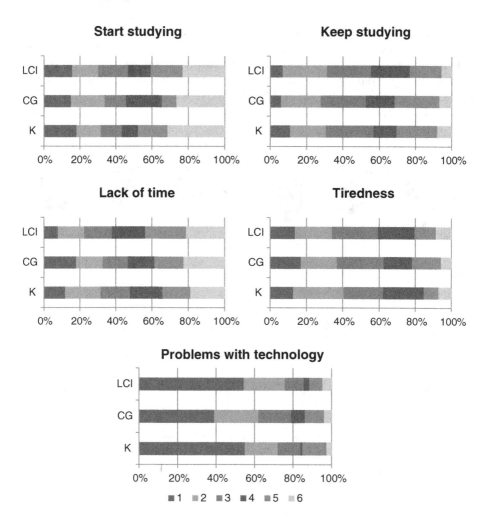

Fig. 5. Learners' problems in the course of study

second criterion, i.e. keep studying, when respondents in the K group reached the highest score of 61 %, followed by the LCI group (55 %) and CG group (53 %).

Approximately 20 % in each group suffered from lack of time for studying (61 % in the LCI group, 54 % in the CG group and 52 % in the K group.

What is rather surprising students did not feel so much tired as could be expected when studying in the combined, i.e. part-time form, after a full-time job. Fifty-nine per cent of respondents in the LCI group had no or slight problems with tiredness before or within learning, and even the higher scores were reached in other groups (63 % both in the CG and K group).

Most respondents did not have substantial problems with technology (85 % in the LCI group, 79 % in the CG group and 84 % in the K group). To sum up these findings (Fig. 6), learners expressed their positive approach to studying online and satisfaction with the course of study. Minimal differences were detected between the groups; the lowest number of dissatisfied students was in the CG group (5 %), were all types of study materials were provided without individual preference recommendation. Hardly any crucial problems appeared which could indicate that neither the online learning (i.e. ICT-supported instruction), nor the entire learning environment built any limits and restrictions to students in the process of learning, and most of them expressed their preference of e-learning to the traditional face-to-face way of teaching/learning. As the satisfaction rate was high, further statistical processing of these data was not applied.

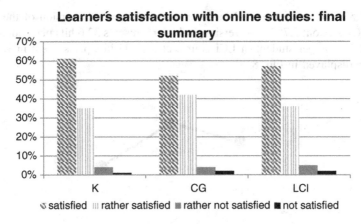

Fig. 6. Learners' satisfaction with online studying

Despite all the possible and real problems the course participants had during the course of study, approximately 80 % would take another course (other courses) within their university study and prefer online learning to traditional face-to-face approach (Fig. 7).

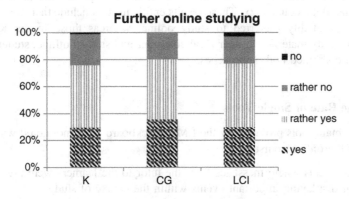

Fig. 7. Further online studying

5 Tracking Learners Activities Within Individualized Learning

So that the above presented findings were supported by real data, learners' activities in all three versions of the online course were tracked by the LMS. Two criteria were considered: (1) the visit rate to the course and (2) the usage of single tools.

5.1 Visit Rate to the Online Course

Totally 12,576 hits were detected in the course (4,271 in CG version of the course; 5,543 in LCI version; 2,762 in K version), which represents 32.6 hits per student in CG course; 42.32 hits per student in LCI course and 20.93 hits per student in K course. Results are displayed in Fig. 8.

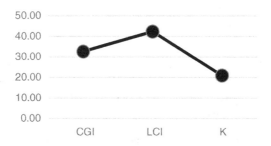

Fig. 8. Visit rate to the three versions of the online course (hits per student)

The data show the visit rate differs substantially in the three versions of the course. Whereas the lowest frequency (21 × per the whole study period) was detected in the K version of the course where the process of instruction reflected teacher's style, in the individualized LCI version the frequency was twice higher (42x); in the CG version where all types of study materials were provided to the learners and the process of selection was the matter of their individual decision, the frequency was in-between, i.e. students entered the course 32x. These results entitle us to conclude that students in the LCI version probably preferred to study online, whereas those in the K version downloaded study materials to their local computer and studied offline; students in the CG version combined both approaches.

5.2 Usage Rate of Single Tools

Totally five main tools provided by the LMS Blackboard were monitored within three versions of the online course:

– Announcements where messages from the tutor to the learners were presented;
– Calendar displaying important events within the course of study;

- Learning content where all study materials were listed in individualized or non-individualized order;
- Discussions running between tutor-learner or learner-learner/s and
- My grades displaying test scores each student reached during the study etc.

The usage of single tools by learners is displayed in Table 7 and Fig. 9.

Table 7. Usage of single tool (in per cent)

	CG	LCI	K
Announcements	1.90	0.76	3.59
Calendar	0.16	0.05	0.22
Content	88.78	96.36	89.48
Discussion	8.71	5.39	2.38
My grades	0.44	0.44	0.44

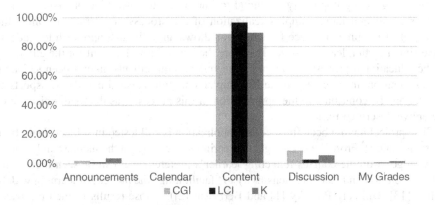

Fig. 9. Usage of single tools (in per cent)

All monitored tools were used to similar extend in all three versions of the online course.

As expected, it is clearly seen the *Learning content* was the most frequently accessed tool within all versions of the online course (96.36 % of learners used this tool in the LCI version, followed by 89.48 % in the K version, and whereas the learning content was the least frequently accessed by learners in the CG version). This result proved previous interpretation that students preferred the work online. Despite the frequency is high, the question appears how the learners could study without accessing the study materials. The interpretation could be that the missing part of learners up to 100 % are those who did not finish the online course.

From other tools *Discussion* was the most frequently used by learners studying in the CG version of the course (8.71 %). A reason might be that having all types of study materials available and their choice was the matter of individual decision, they longed for discussions, experience and feedback from other learners. This group was followed by K group, whose studies reflected teachers preferences and 5.39 % of learners were involved in discussions; and, only 2.38 % of students in the LCI version of the course participated in discussions. This rate was even lower than with *Announcements* in the K group, which reached 3.59 %. It is higher compared to 1.9 % in the CG group; and, only 0.76 % of students in the LCI version of the course read tutor's announcements. Remaining two tools (My grades, Calendar) did not exceeded the 1 % value (*My grades*: 0.44 % in CG and LCI versions of the course, 1.33 % in K group; *Calendar*: 0.16 % in CG, 0.05 in LCI, 0.22 % in K group). These results indicate minimum usage of these tools which can be interpreted students probably did not consider them important for their learning.

6 Discussion and Conclusion

Current orientation of university education, which is changing under the influence of latest technology development and new key competences, can be researched from various, different points of view. The technology enhanced learning has been spreading because of growing popularity of digital technologies in general. Another reason is it enables easier and more complex realization of the process of instruction, offers the choice of place, time and pace for studying, allows an individual approach to students preferring a certain learning style. These are the key values important for the efficiency of the educational process. Material and technical requirements having been satisfied, strong attention must be paid to the ICT implementation reflected in didactic aspects of instruction. To contribute to the optimization of this process was the main objective of the above described research.

The project having been finished, the-application is still tested in subjects Data-base Systems I and II. From the pedagogical experiment focusing on the increase in learners' knowledge in online courses reflecting learner's preferences it can be seen there is no definite solution and students' sensitivity to "facilitating" the process of learning widely differs [13]. Unlikely Prensky [1] and Berry (in [2]), whose results formed the background of our project, our results proved most students of IT study programs (Applied Informatics, Information Management) were flexible in learning to such extent they reached the same results either the process of instruction reflected their learning preferences, or not. Gulbahar and Alper developed the e-learning style scale, collecting feedback from 2,722 students of distance study programmes. Starting with 56 items categorized in eight groups, they finally defined 38 criteria structured in seven groups. [14] As our research did not prove statistically significant differences in learners' knowledge in the experimental and control groups, we are going to use the Gulbahar and Alper's scale to be applied in the future research of the technology enhanced instruction. Two decades ago Mehrlinger (in [15]) emphasized that a variety of learning styles influenced the teacher-designer's teaching methods and choice of media in a given course/lesson and predicted that technology of the future would be more integrated, interactive, and intelligent. Integration continued to escalate through the development of advanced multimedia systems and interactivity occurred with increased distance learning and Internet interaction, followed by individualized knowledge addressing the learning styles of each student. That has been imperative for teachers to keep abreast of technological changes to empower their students.

As stated by the EC-TEL 2013 conference (http://ectel2013.cs.ucy.ac.cy/), there is no doubt that technology enhanced learning has created enormous changes in educational institutions of all levels and at the workplaces. However, these innovations have tended to be unsustainable – they need a high degree of effort to be sustained, i.e. mainly funded. At the same time, the technology (mobile and social information and communication technologies) makes impact on everything and everybody around. And, above all, most of educational institutions have taken these technologies up in a systematic way to include them into their learning strategy, sustain them and develop by reflecting feedback provided by research activities.

The ICT contribution to the individualization of the process of instruction supported by e-application was researched by the method of pedagogical experiment. Unfortunately, neither the research results, nor learners' evaluation proved our expectations that the reflection of individual learning style might be the means which (if applied in the didactic and sensitive manner) could help substantially within the process of online learning. This result was surprising because the learning style reflection had been understood to be a powerful factor providing strong impact on the process of learning, and statistically significant increase in knowledge of the LCI course participants was expected.

There might be several reasons how to interpret the results.

First, neither strong, nor marginal preferences were discovered in patterns within the sample group which could produce statistically significant differences. We agree with e.g. Gregorc [16] or Mitchell [17] saying that not tailoring the process of instruction to learners' individual preferences results in increase the knowledge but they consider developing new learning strategies to be more contributive to the learner. Thus the current research question states as follows: *Is it really worth dealing with learning styles if the pedagogical experiment did not prove any increase in knowledge?* Despite the unexpected research result, our answer is 'yes', as all three groups of learners studying in three different versions of the online course declared their satisfaction with the process of instruction they went through, including those in groups where individual learning preferences were not reflected.

Second, as mentioned above, there exist some researches (and researchers) that reject the theory of learning styles resulting in the individually tailored process of instruction. The proposal might be to provide the individualized process of instruction (a) to learners showing very strong preferences in one learning style, and help them develop other strategies and approaches; (b) attract attention and show those who have very weak preferences and are able to study efficiently using any strategy that there exist approaches and methods which might suit them better, which finally can increase their motivation in learning, make the process more interesting for them, which is not of little importance.

Third, there could be several other reasons why the expectations and hypotheses were not verified, both on the researchers' and learners' side. In further research activities other approaches to running the process of instruction reflecting individual learning styles can be tested, i.e. tutor's role as a facilitator could be strengthened and emphasized so that learners feel and study in a more friendly environment, being provided wider technical and didactic support; learner's experience in online learning developed in this course could be extended and many other measures could be taken.

On the learners' side the skill of independent work and study must be supported and gradually developed, as online learning has become standard not only in the tertiary education but particularly in lifelong learning.

One of the project outcomes - the e-application generating the learning content in adequate order has been designed and can be used as freeware on request but no increase in learners' knowledge was discovered. What has been appreciated is the learners' positive approach to online learning which was expressed by their approach to further learning in online courses. Despite this factor did not belong to the primary or crucial ones, it can be considered a positive side contribution without hesitation.

Thus it can be concluded that despite the contribution of the learning style theory to the online learning process was not proved within this project, no decrease in learners knowledge was discovered in comparison to the traditionally led process of learning which follows teacher's style of instruction. As mentioned in the first chapter, the time came to deal with didactic aspects of ICT implementation into the process of instruction. And, the final question still exists: What else can be done to make the process of learning easier? Following the Felder's multistyle approach [18] we would recommend to use a wide range of methods, strategies and approaches which have been successfully applied in the face-to-face form of instruction for ages and use them under the conditions of e-learning. The Bloom's digital taxonomy [19] introduced by Churches might be one of the tools.

"Only providing technologies does not change the situation much, but it can start new activities and approaches. Bringing computers to schools is less important than provide teachers with new ideas. Technologies do not aim at removing traditional educational methods and forms. The new technologies do not automatically bring positive changes into the process of instruction. But they may contribute to increasing its effectiveness, under some conditions" [20].

Acknowledgment. This paper is supported by the SPEV Project N. 2110.

References

1. Prensky, M.: Digital natives, digital immigrants. http://www.marcprensky.com/writing/Prensky%20-%20Digital%20Natives%20Digital%20Immigrants%20-%20Part1.pdf (2001)
2. Prensky, M.: Sapiens digital: from digital immigrants and digital natives to digital wisdom innovate. http://www.marcprensky.com/writing/Prensky%20-%20Digital%20Natives,%20Digital%20Immigrants%20-%20Part2.pdf (2009)
3. Felder, R.M., Silverman, L.K.: Learning/teaching styles in engineering education. J. Eng. Educ. **78**(8), 674–681 (1998)
4. Gregorc, A.F.: Learning/teaching styles: potent forces behind them. Educ. Leadersh **36**, 234–2387 (1979)
5. Mitchell, D.P.: Learning style: a critical analysis of the concept and its assessment. Kogan Page, London (1994)
6. Coffield, F., et al.: Learning styles and pedagogy in post-16 learning. A systematic and critical review. Newcatle University Report on Learning Styles (2004)
7. Honey, P.: Learning styles - the key to personalised e-learning. http://www.bbmatters.net/bb_matters.../Learning%20styles_peter%20honey.pdf (2010)

8. Honey, P., et al.: Attitudes to e-learning: A national survey 2000 undertaken by the campaign for learning. Southgate Publishers, Sandford (2000)
9. Honey, P.: E-learning: a performance appraisal and some suggestions for improvement. http://qou.edu/arabic/research/Program/e-learningResearchs/e-learningPerformance.pdf (2010)
10. Johnston, C.A.: Unlocking the will to learn. Corwin Press Inc, Thousand Oaks (1996)
11. Kolb, D.A.: Experiential learning: experience as the source of learning and development. Prentice-Hall, Englewoods Cliffs (1984)
12. Honey, P., Mumford, A.: Using Your Learning Styles. Peter Honey Publications, Maidenhead (2002)
13. Šimonová, I., Poulová, P.: Learning Style Reflection Within Tertiary e-Education. WAMAK, Hradec Kralove (2012)
14. Gulbahar, Y., Alper, A.: Development of e-learning styles scale for electronic environments. Educ. Sci. **39**(171), 421–435 (2014)
15. Rogers, P.L.: Designing Instruction for Technology-Enhanced Learning. Idea Group Publishing, London (2002)
16. Gregorc, A.: Gregorc's mind styles. http://www.colorfulleadership.info/papers/concrete.htm (2004)
17. Mitchell, D.P., et al.: Learning Style: A Critical Analysis of the Concept and Its Assessment. Kogan Page, London (2004)
18. Felder, R.M.: Are Learning Styles Invalid? (Hint: No!). http://www4.ncsu.edu/unity/lockers/users/f/felder/public/Papers/LS_Validity(On-Course).pdf (2010)
19. Churches, A.: Bloom's digital taxonomy. http://edorigami.wikispaces.com/Bloom%27s+Digital+Taxonomy (2010)
20. Venezky, R.L., Davis, S.C.: Quo vademus?: the transformation of schooling in a networked world: (case study report): OECD/CERI version 8c. OECD. http://www.oecd.org (2002)

GLIO: A New Method for Grouping Like-Minded Users

Soufiene Jaffali[✉], Hanen Ameur, Salma Jamoussi,
and Abdelmajid Ben Hamadou

MIRACL Laboratory Higher Institute of Computer Science and Multimedia,
University of Sfax, BP 1030, Sfax, Tunisia
jaffali.soufiene@gmail.com

Abstract. Grouping like-minded users is one of the emerging problems
in Social Network Analysis. Indeed, it gives a good idea about group
formation and social network evolution. Also, it explains various social
phenomena and leads to many applications, such as friends suggestion
and collaborative filtering. In this paper, we introduce a novel unsupervised method for grouping like-minded users within social networks.
Such a method detects groups of users sharing the same interest centers and having similar opinions. In fact, the proposed method is based
on extracting the interest centers and retrieving the polarities from the
user's textual posts. We validate our results by employing multiple clustering evaluation measures (recall, precision, F-score and Rand-Index).
We compare our algorithm to a number of other clustering algorithms
and opinion detection API. Results prove that the algorithm presented
is efficient.

Keywords: Social network · Like-minded users · Interest centers · Sentiment analysis

1 Introduction

Identifying and understanding groups of users sharing similar interests are emergent tasks that allow us to uncover organizational principles in networks. To
detect communities in social networks, there are two possible sources of information one can use: the network structure, and the features and attributes of
nodes. Most of works concerned with this issue deal with it as a graph distribution problem, in which the users are represented by nodes and the relationships
between them by edges [34]. These relationships are generally explicit friendship
links (*"friend"* on Facebook, *"follower/followee"* on Twitter, etc.). According to
the big tail distribution of social networks [33], most of the social media users
have only a few links. Therefore, it is hard to find like-minded people who are
several steps away from each other within the same social network by considering
only explicit links. In addition, regarding the huge number of social network users
(over 645,750,000 active registered Twitter users according to Statistic Brain[1]),

---

[1] http://www.statisticbrain.com/twitter-statistics.

© Springer-Verlag Berlin Heidelberg 2015
N.T. Nguyen (Ed.): Transactions on CCI XVIII, LNCS 9240, pp. 44–66, 2015.
DOI: 10.1007/978-3-662-48145-5_3

mining only explicit relations within the network do not provide a complete vision. This implies a limitation of the link based approaches.

In this study, we propose to group the users sharing the same interests by analyzing their textual posts. The main goal is to retrieve the interest centers from the users' posts and, then, to group those having the same interests. At this stage, we can find, in a given group, users having opposite opinions about the same subject. So, they cannot be considered as like-minded users. To overcome this problem, we add a sentiment-analysis to know whether the user has a positive or a negative opinion about the interest center. Thus, we obtain two sub-groups by interest center. Grouping like-minded people based on their interest centers and their polarities, is a very interesting task. Indeed, it improves the quality of recommendation and social marketing systems, and leads to many applications like poll systems and familiar stranger recommendation.

The rest of the paper is organized as follows. In Sect. 2, we give an overview of the related work. In Sect. 3, we present our proposed approach to group likeminded users. We describe our experiments in Sect. 4. The results are in Sect. 5. We end up with the conclusions and the future work in Sect. 6.

2   Related Work

2.1   Community Detection

In the literature, the majority of work dealing with the subject of user's classification and community detection is based on link information [12]. The *divisive algorithms* are one of the most known link-based approaches. The main idea of those approaches is to remove iteratively, from a given graph, edges that connect vertices of different communities, in order to disconnect clusters from each other [15]. The *modularity-based methods* are inspired from the modularity function, originally introduced to define a stopping criterion for the divisive algorithm of Girvan and Newman [34]. This function has rapidly become an essential element of many clustering methods. The main idea of *modularity-based methods* is to join vertices when the merging increases the modularity. The *dynamic algorithms* are also largely used for community detection. Those methods employ processes running on the graph, focusing on spin-spin interactions, random walks and synchronization [58]. We can also cite the *statistical inference based methods*. The latters aim at deducing properties of data sets, starting from a set of observations and model hypotheses [18]. Mining only explicit links in a social network is very important to study the social network evolution. However, as the link information informs only about the explicit relations, and given the big tail distribution of user links and the huge size of social networks, mining only link information does not provide a complete vision. Also, such community detection methods are not optimal for collaborative filtering, because we have not necessarily the same preferences with our social network friends. In addition, suggesting friends based on the link information can mostly recommend people you may already know.

Moreover, many kinds of information are used to retrieve significant communities, such as the mutual awareness [28], comments and like actions [38]. [3] used the link patterns to measure the degree of interaction between two political communities. Also, tags are deeply used to construct the user profiles [13], and to classify the interest centers [26]. [1] created tag communities using the Principal Component Analysis (PCA) and assign the users to the closest communities. [53] connected the like-minded users using the tag network inference. Given the fact that some users do not employ tags in their posts and that the same subject can be described by more than one tag (e.g. "#Tnelec", "#MMM", "#ISIE" and "#tunisiaelections" describing the Tunisian elections 2014), the use of tags for community detection may not succeed or yield to unoptimized results. Therefore, we suggest using textual posts which are rich of information to retrieve the interest centers within the social network, and then, using the retrieved centers in order to group the users into communities. In the literature, just a few works deal with extracting social relations between individuals from text [32].

For clustering purposes, some proposed approaches combine text and link information. Aggarwal et al. relie on K-means algorithm for the text attributes and Bayesian probability estimations for the side attributes in [5]. As the approach proposed in [5] is general, the side attributes can be any kind of information rather than link-based information. Angelova et al. proposed two methods of clustering in [7]. The first one uses the link information to adjust the weights of terms. In this method, the common term between a document and its neighbors has more importance in the clustering process. The second method is a graph-based approach. To perform the clustering, this approach relies on a Markov Random Field (MRF) technique. In the context of community detection, the text information in social graph can be attached to either to nodes [56] or to edges [41].

When the communication between different nodes is extensive, which is the case of online chat networks, it is assumed that the text content in the network is attached to the edges. In [41], a matrix-factorization methodology is used in order to jointly model the content and structure to detect communities. On the other hand, the node-based approach is generally the most common. In [56], a node-based approach is proposed, consisting of a combining link and content information. The authors in [56] use a conditional model to analyze links, and a discriminative content model to present content attributes. Then, combines the two models into a unified framework and proposes a novel two-stage optimization algorithm for the maximum likelihood inference. A detailed survey of text clustering can be found in [4].

Topic modeling is widely used in the context of community detection. In this context, we attempt to model the probability of a document belonging to a particular community. The Latent Dirichlet Allocation (LDA) and the probabilistic Latent Semantic Analysis (pLSA) are largely implemented to generate the subject models being used to regroup tweets [20]. Similarly, [45] applied LDA to identify the subjects of discussion based on the interactions between users. These subjects are used to create the communities in a second stage. Using LDA,

[20] extracted the subjects from the published tweets to build a subject-based model. [49] proposed the scalable multi-stage clustering algorithm (SMSC) in order to categorize tweets. The SMSC algorithm had been tested on a collection of tweets and presented high performances.

In our case, we propose an algorithm based on the PCA method to group like-minded people using their textual posts. Our algorithm differs from the classic text based community extraction methods, since it allows generating signed communities (positive and negative) according to users polarities. The closest work to the proposed approach is presented in [2], in which, the authors try to identify the sub-groups in online discussions. The common principle of these works is to use the text mining tools to locate the texts comprising opinions, and to specify their targets. According to the opinion targets, the researchers gather the similar texts. By using a grouping algorithm, they subdivide the users in sub-groups. The major problem of such user classification is that it takes into account only text with information about polarity, and so a great part of information is ignored. Thus, we suggest retrieving the interest centers in the first step, then, mining the users polarities according to the retrieved interest centers.

2.2   Sentiment Analysis

Within a group of users talking about the same interests center (same subject), we can find some users agree with this subject and some others disagree. Therefore, it is very interesting to distinguish between those who have the positive and the negative opinion towards a specific subject (representing an interest center). For this reason, we perform an additional task for sentiment analysis. Generally, the goal of this task is to identify and classify the subjective texts into two classes (positive and negative) or three classes (positive, negative and neutral). In the literature, a fairly significant number of studies have tried to address the problem of automatic sentiment analysis using machine learning or other techniques. In fact, there are three major approaches for sentiment analysis: namely, linguistic, statistical and hybrid.

The linguistic approach is based on the use of a set of classified words (called sentiment lexicon or dictionary) to classify, thereafter, a larger text. The step of lexicon construction is then essential in any application based on this approach. Indeed, this construction step can be performed with four kinds of techniques: The first technique is manual by experts, e.g. [55]. The second is based on external lexical resources (such as General Inquirer, WordNet, etc.) using the semantic relations (synonymy and antonymy) that they manage, e.g. [11]. The third one is based on the information present in the text using coordinating conjunctions (and, but, etc.) [10], co-occurrence of words [51], or emotion symbols (emoticons, Acronyms and exclamation words) [6,23,47]. The fourth technique consists of combining the two latter techniques, i.e. the use of corpus information and external sentiment resource (dictionary). For example, [37] have proposed a method based on using corpus (tweets) and ANEW dictionary in order to construct an affective lexicon in French. Their procedure is to divide the corpus into two sets of messages by using the emoticons as an indicator of general text sentiment.

In the area of sentiment classification, the most used statistical methods are the machine learning methods (Supervised methods). This kind of method requires a training set (manually labeled data) to classify other unlabeled data. The popular algorithms in machine learning used to train the sentiment classifier of words and sentences are SVM (Support Vector Machines) [21], Naive Bayes [21,35]. In order to perform these methods, it is necessary to represent numerically the data (words and phrases). In the literature, two types of representation are used in data mining and which are also used in sentiment mining. (i) The first representation is the vector representation "bag of words" (binary vector [35], Frequency vector [40,57], TF-IDF vector [40]) and (ii) the second representation is sequential representation "n-gram", such as [37].

In the hybrid approach, linguistic techniques and statistical ones are combined in order to improve the classification results. This hybridization can be done by employing the linguistic techniques within the statistical approach [54], and vice-versa [52], to improve the results.

3   Grouping Like-Minded Users

Given a group of users E, we intend to find the optimal distribution of the group E in K clusters. The optimal distribution maximizes the correlation between the intra-group users and minimizes the similarity between the intergroup users. In our approach, the grouping is based on the content of the messages. The main goal here is to find the latent centers of interest around which the input data are concentrated. Then, we calculate the distances between the users and the centers that we found. Ultimately, we gather the users according to the distances which separate them from the interest centers. Thus, the users close to the same center belong to the same group. Each group is then divided into two sub-groups according to the polarity of the users. The five principal steps of our GLIO algorithm (Grouping Like-minded users based on Interest-centers and Opinions) are hereafter detailed. Figure 1 presents the proposed method.

3.1   Text Preprocessing

This phase consists of filtering the raw data and displaying them in a new representation adequate for the following steps. In the present work, we deal with the Twitter text messages known as "tweets". Those latters are limited to 140 characters, allowing users to share their opinions. Tweets may contain certain meta-data such as words prefixed with '#' (known as "hashtag") to describe their subjects, with '@' to mention or answer other user, and with 'RT' to 'retweet' (Republish) another tweet. Tweets are neither structured nor written in a formal language which may make their exploitation very difficult. To remedy this problem, we start by eliminating stop words (personal pronouns, prepositions, etc.) and converting all upper-case letters to lower-case ones. Next, we keep only words appearing a number of times more than a prefixed threshold value. Thus, we obtain the bag of words $T = \{t_1, t_2, ..., t_m\}$. Then, we group the tweets

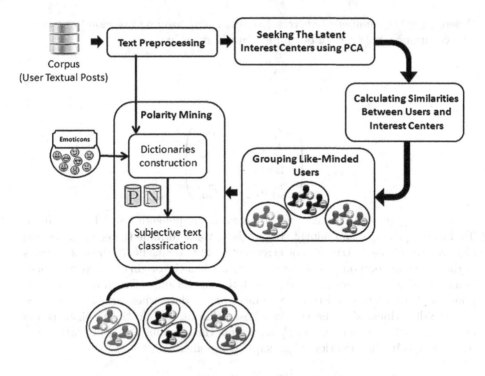

Fig. 1. Proposed method for grouping like-minded users

of each user. Using the bag of words approach, we represent each user by the occurrence-appearance values of the m most frequent words in our data. Thus, we obtain a representative vector of each user. Finally, we combine these vectors to form a matrix *users × terms*.

3.2 Seeking the Latent Centers of Interest

In this step, we seek for the latent interest centers within the input data. To do this, we use the Principal Component Analysis (PCA) method [39]. The latter proved its effectiveness in the theme detection task and textual documents clustering [24]. PCA is generally used to reduce the data space or to determine the axes where data are concentrated as it is the case of our study.

To retrieve the axes of concentration of data from a given cloud of points, the PCA method consists in seeking the axes (directions) of our cloud of points where the variance is maximal. Therefore, we calculate the square matrix of covariance of input data and its eigenvalues and eigenvectors. To do this, we start by calculating the average $\bar{p}$ (the center of our cloud of points), using the following function:

$$\bar{p} = \frac{1}{m} \sum_{k=1}^{m} p_k \tag{1}$$

Where m is the number of vectors $\{p_1, p_2, ..., p_m\}$ forming the cloud of points. The covariance matrix C of our input data is given by:

$$C = \frac{1}{m-1} \sum_{k=1}^{m} (p_k - \bar{p})(p_k - \bar{p})^T \tag{2}$$

$$= \begin{pmatrix} \sigma_{11}^2 & \sigma_{12}^2 & \cdots & \sigma_{1m}^2 \\ \sigma_{21}^2 & \sigma_{22}^2 & \cdots & \sigma_{2m}^2 \\ \vdots & \vdots & \ddots & \vdots \\ \sigma_{m1}^2 & \sigma_{m2}^2 & \cdots & \sigma_{mm}^2 \end{pmatrix} \tag{3}$$

Where σ_{ij}^2 presents the covariance between components i and j in our data. To find the principal components of our data, we determine the eigenvalues and eigenvectors of the matrix C. The eigenvectors of C define in $\Re^n$ the orientations of the principal components of our input data when the origin of the vector space is moved to $\bar{p}$. The eigenvalues represent the significance of each of these components. They correspond to the variances of the data when projected on each of these directions. Z is the matrix whose columns contain the m eigenvectors of C. It's a rotation matrix and P_k' represents the stimulus after translation and rotation axis in the direction of principal components.

$$P_k' = Z^{-1}(p_k - \bar{p}) = Z^T(p_k - \bar{p}) \tag{4}$$

If we calculate the covariance matrix C' of P_k', we obtain a diagonal matrix Λ whose elements correspond to the eigenvalues of C:

$$\Lambda = \begin{pmatrix} \lambda_1 & 0 & \cdots & 0 \\ 0 & \lambda_2 & \cdots & 0 \\ \vdots & \vdots & \ddots & \vdots \\ 0 & 0 & \cdots & \lambda_m \end{pmatrix} \tag{5}$$

In our work, the data are the terms used by the users in their tweets. Consequently, the principal components are the axis around which the terms are concentrated. In other words, these components are the latent interest centers in the tweet collection. Let T be the bag of words used in the users posts, and $U = \{U_1, U_2, ..., U_i, ..., U_n\}$ the occurrence matrix, with U_i is the vector representing the $user_i$, and n is the number of users. Each user's vector is of the form $U_i = \{o_{i1}, o_{i2}, ..., o_{ih}, ..., o_{im}\}$, with o_{ih} is the occurrence number of the term t_h ($t_h \in T$) in the posts of the user U_i. We calculate the covariance matrix of U and its eigenvalues and eigenvectors. The obtained eigenvectors present the latent interest centers within the users' posts. Each interests center is of the form $C_j = \{c_{j1}, c_{j2}, .., c_{jl}, .., c_{jm}\}$ with c_{jl} is the weight of the term l within the component j. The terms having the highest weights in C_j are those reflecting its subject [24]. We use the eigenvalues to determine the number of clusters. This issue will be explained in Sect. 3.4.

Table 1. Example of occurrence matrix

	Word 1	Word 2	Word 3	Word 4	Word 5	Word 6
User 1	1	2	1	0	0	0
User 2	2	1	1	0	0	0
User 3	0	0	0	1	3	2
User 4	0	0	0	2	2	1
User 5	2	2	0	0	0	0
User 6	1	3	1	0	0	0
User 7	0	0	0	2	1	1
User 8	0	0	0	1	2	2

To illustrate how we seek the latent interest centers, we introduce the following example. Let U be the occurrence matrix presented in Table 1. Where, each line presents one user. The first user uses once the first and the third words, twice the second word; however, he does not employ neither the fourth, the fifth nor the sixth words.

We calculate the eigenvalues and eigenvectors of the covariance matrix of U. The obtained eigenvectors represent the latent interest centers, we group those eigenvectors in the matrix C.

$$C = \begin{pmatrix} -3.6E-01 & 9.9E-02 & 6.9E-01 & -2.6E-01 & 2.4E-01 & 4.8E-01 \\ -5.1E-01 & -4.6E-01 & -5.5E-01 & -3.6E-01 & 7.8E-02 & 2.6E-01 \\ -1.8E-01 & -8.6E-02 & -1.0E-01 & 8.5E-01 & 2.8E-01 & 3.5E-01 \\ 3.5E-01 & 5.8E-01 & -4.1E-01 & -2.0E-01 & 2.3E-01 & 5.1E-01 \\ 5.3E-01 & -5.2E-01 & 1.0E-01 & -1.2E-01 & 6.4E-01 & -3.1E-02 \\ 3.9E-01 & -3.8E-01 & 1.0E-01 & 3.4E-02 & -6.2E-01 & 5.4E-01 \end{pmatrix}$$

Each column in C represents an eigenvector, and so an interest center. The values in a given eigenvector represent the weight of words in the corresponding interest center. The obtained eigenvalues are:

$$\lambda = \{4.48, 0.43, 0.34, 0.13, 0.06, 0.00\}$$

3.3 Assigning Users to Their Interest Centers

This phase consists in defining a new representation of users by taking into account the interest centers identified in the previous step. Given that the users who share similar interests are close to the same principal components, we represent each user by his distance from all the centers of interest. To do this, we calculate the distances between the user's word occurrence vector $U_i = \{o_{i1}, o_{i2}, ..., o_{ih}, ..., o_{im}\}$ (calculated in the first step) and each eigenvector $C_j = \{c_{j1}, c_{j2}, .., c_{jl}, .., c_{jm}\}$ (principal component). In this study, we use the cosine as a metric to measure the deviation between vectors. As we The similarity between the user U_i and the interest center C_j is then given by this formulation:

$$Similarity(U_i, C_j) = \frac{U_i \cdot C_j}{\|U_i\|\|C_j\|} \tag{6}$$

Thus, each user U_i is, then, represented by a vector U_i', with:

$$U_i' = \{S_{i1}, S_{i2}, .., S_{ik}, .., S_{im}\}$$

Where S_{ik} is the similarity between the user U_i and the interest center C_k.

Considering the previous example, after calculating the similarities between users' vectors and the calculated interest centers, we obtain the new representation of the users U'.

$$U' = \begin{pmatrix} -0.64 & -0.37 & -0.20 & -0.05 & 0.27 & 0.55 \\ -0.58 & -0.14 & 0.30 & -0.015 & 0.34 & 0.65 \\ 0.73 & -0.46 & 0.02 & -0.13 & 0.24 & 0.40 \\ 0.72 & -0.08 & -0.17 & -0.20 & 0.37 & 0.50 \\ -0.62 & -0.26 & 0.10 & -0.44 & 0.22 & 0.52 \\ -0.63 & -0.41 & -0.32 & -0.14 & 0.23 & 0.49 \\ 0.66 & 0.10 & -0.25 & -0.20 & 0.19 & 0.63 \\ 0.73 & -0.41 & 0.00 & -0.12 & 0.09 & 0.51 \end{pmatrix}$$

3.4 Users Clustering

In this step, we employ the K-Means algorithm to regroup the like-minded users in categories. Indeed, K-Means is a grouping algorithm that classifies objects in K groups by taking into account their attribute values. It is known by its simplicity and effectiveness. As inputs, K-Means takes a set of data D and the number of groups to be identified K. Then, K centers of gravity are fixed randomly. In the second step, each object of the data set is allocated to the cluster C having the closest center. After the allocation of all objects in clusters, the centers are recomputed by taking into account the allocated objects. The algorithm reaches its end when no change is observed and, thus, K groups are obtained as output.

The main problem of the K-Means algorithm lies in the difficulty in finding the number of clusters K, which is an NP-complete problem [31]. Most of the work using the K-Means algorithm either set the value of K arbitrary or vary it empirically, which is the case of [48,49].

To overcome this problem, we use the eigenvalues calculated in the second step. Since the principal components correspond to the latent interest centers in the group of users, the knowledge of the component number is related to the knowledge of the number of interest centers and whence the clusters.

In the literature, many methods are suggested to determine the component number to be considered [9,25]. In this work, we adopt the Scree Test Acceleration Factor proposed in [42] as a non-graphical solution of the method of Cattel [9]. In fact, this solution consists in considering the number of principal components which precede the coordinate where the *acceleration factor* is maximized. This latter is indicated by the abrupt change in the slope of a function f curve. With f is a function that passes through all the eigenvalues. The acceleration factor can be given for any i (between 2 and $p-1$ with p is the number of eigenvalues) by the second derivative $f''(i)$. The simplified function of the second derivative can be written as follows:

$$f''(i) = f(i+1) - 2 * f(i) - f(i-1) \tag{7}$$

The considered components must also have eigenvalues higher than 1 or than the Location Statistic criterion LS (generally the median, the average or one of the 0.05, 0.50 or 0.95 centile). Thus, we obtain the following system:

$$\begin{cases} n_{af} = Count[(\lambda_i \geq 1 \ \& \ i \leq k) & with \ k \equiv arg \ max_j(af_j)] \\ or \\ n_{af} = Count[(\lambda_i \geq LS_i \ \& \ i \leq k) \ with \ k \equiv arg \ max_j(af_j)] \end{cases} \tag{8}$$

With λ_i is the ith eigenvalue, and af is the acceleration factor. Figure 2 represents the eigenvalues calculated for the previous example. According to Cattel [9], the screw seems to appear at the second eigenvalue. To find the number of clusters using the Scree Test Acceleration Factor, we start by calculating the LS criterion. Thus, we calculate the median of the eigenvalues, we obtain $LS = 0,23$. Then, we use the second derivative to calculate the af for the eigenvalues between 2 and 5 (because we have 6 eigenvalues). We obtain the following acceleration factors:

$$af = \{3.2694, -0.3779, 0.0009, -0.0083\}$$

We note that the second eigenvalue has the maximum acceleration factor. So, the $arg \ max_j(af_j)$ in the Eq. (8) is equal to 2. According to the Eq. (8), to calculate the number of clusters, we use the following equation:

$$n_{af} = Count[(\lambda_i \geq 0,23 \ \& \ i \leq 2)$$

We obtain a number of clusters equals to 2, which is the same number of clusters given by the Cattel's method.

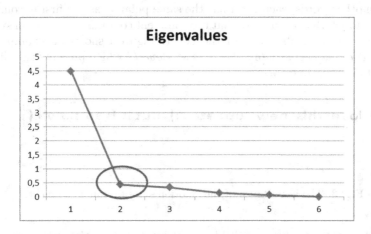

Fig. 2. Obtained curve of the eigenvalues of the example matrix U

As the number of clusters in our example is equal to 2, we use the K-Means algorithm to classify users into two clusters. We use the matrix U', calculated in the third step, as input of the K-Means algorithm. We obtain the following distribution:

$$users\ distribution\ in\ clusters = \{1, 1, 2, 2, 1, 1, 2, 2\}$$

Thus, the first, second, fifth and the sixth users belong to the same group. In other hand, the users 3, 4, 7 and 8 form the second group. Considering the matrix U, we note that the users of the same group use the same words in their posts, and so they have the same interest centers.

3.5   Polarity Mining

At this stage, we obtained groups of like-minded users, which talk about the same subject "interests center". In order to determine the polarity of users' sentiment (positive or negative) toward their interest centers, we add the sentiment analysis module (polarity mining step). To do so, we use the method described in [6]. In fact, this method consists of determining the sentimental orientation "positive or negative valence" of a Facebook comment by using the linguistic approach. It is depicted in two essential tasks; (i) the first is to construct a lexicon containing words expressing sentiments which are categorized into two classes (positive and negative). (ii) The second aims to determine the overall polarity of subjective texts. This method creates dynamically the positive and negative dictionaries covering the majority of sentiment words lexicon, based on the presence of emotion symbols in the comments of the corpus. Thus, they predicted the positive or negative polarities of comments are predicted using these prepared sentiment dictionaries.

Furthermore, throughout the dictionaries construction step, we focused on the hypothesis that the emotion symbols (emoticons, Acronyms and exclamation words) reflect the sentiment expressed by the words that precede them in the tweets. In other words, each word has the same polarity as the first encountered emotion symbol. But, in the case that the comment contains one emotion symbol, the latter affects on all words of the tweet. Figure 3 shows an example of a tweet *"I love this new update :) but it is so slow :("* which presents two distinct sentiments in the same tweet.

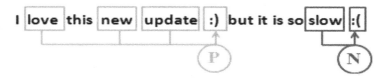

Fig. 3. The importance of the emotion's position in a comment.

In order to create dynamically the sentiment dictionaries corresponding to our corpus, we divided the corpus into two sub-sets of tweets. The first one is based on tweets containing emotion symbols to construct the initial dictionaries; and the second one contains the rest of tweets which haven't emotion symbols for enriching these dictionaries by associating other words to them.

First of all, we begin by identifying two initial dictionaries, one contains words attached to positive sentiments and the other contains words attached to negative sentiments. To do this, we calculate for each lexicon word its positive and negative valences by using the formula 9. Thus, we compare its valences in each of the dictionaries to determine the polarity of every lexicon word. In fact, if it has a positive valence greater than the negative valence, it will be automatically classed as positive and vice versa.

$$valence(w)_j = \frac{frequency(w)_j}{\sum_{i=0}^{n} frequency(w_i)_j} \times 1000 \qquad (9)$$

Where:

- $frequency(w)_j$ is the occurrence number of word w in all segments of a sentence containing an emotional symbol having the polarity j.
- n is the number of words in the dictionary j, with $j \in (Postive, Negative)$.

In order to handle the presence of negation particles (like 'none', 'not', 'neither') in the text, we consider that the polarities of all the words preceded by one of these particles, must be reversed. For this, we also handle the negation particles during the dictionaries construction step. When we calculate the frequency of a word directly preceded by a negation particle, we increment its frequency by 1 in the dictionary that corresponds to the inverse tweet's polarity (the inverse polarity of the emotion symbol).

In order to adjust and correct the valences of the existing words in the dictionaries and add new words to the dictionaries, an enrichment step is achieved. This step is meant to calculate the positive and negative valences of a tweet that does not contain emotion symbols (see Eq. (11)), to increase the positive frequency of the existing words and to add the nonexistent words (new words) in the positive dictionary by the positive percentage of comment (respectively negative), and vice versa. The polarity percentage of a tweet t is calculated using the following formula:

$$percentage_j(t) = \frac{valence(t)_j}{valence(t)_{Positive} + valence(t)_{Negative}} \qquad (10)$$

Where:

$$valence(t)_j = \frac{\sum_{i=0}^{p} frequency(w_i)_j}{\sum_{l=0}^{q} frequency(w_l)_j} \times 1000 \qquad (11)$$

p is the number of tweet's words t existing in the dictionaries, and q is the number of words in the dictionary j. $j \in (Positive, Negative)$.

After the creation of the positive and negative dictionaries, we determine the polarity of each tweet by calculating its positive and negative valences. These valences are calculated as follows:

$$valence_{Pol}(t) = \sum_{i=0}^{p} valence_{Pol}(w_i) \tag{12}$$

Where: p is the number of words in the tweet t.

4   Experimentations

4.1   Baseline

In order to evaluate and position our new approach, we use three reference grouping algorithms; namely, K-Means, LDA and SMSC [49] methods. Based on the *Sentiment140*[2] API we evaluated the implemented sentiment analysis algorithm.

- **K-Means:** As a first reference, we use the classical K-Means [30], one of the most used algorithms for the clustering. Liu in [29] uses K-Means to detect the communities in the networks. We use K-Means to cluster vectors representing users.

- **LDA:** LDA is a generative model that explains a set of observations by unobserved groups that explain the similarity of some parts of the data [8]. It is conceived to analyze the latent thematic structures in large scaled data, including large collections of text or web documents. To implement LDA, we use GibbsLDA++, which is a C/C++ implementation of LDA by using the sampling technique of Gibbs to estimate the parameters and to make the inference.

- **SMSC:** Authors in [49] propose the SMSC algorithm (Scalable Multi-Stage Clustering) for clustering micro-messages. Given a large collection of micro-messages D, and the set of tags T appearing in D, the SMSC algorithm starts by (1) creating a set of virtual documents D' (each $d^t \in D'$ is a concatenation of all micro-messages in D that contain a specific tag $t \in T$). The number of documents in D' is equal to the number of tags in T; (2) In the second step, the SMSC classifies D' messages by applying the K-Means algorithm; (3) Then, it re-transforms each virtual document into its original version by assigning each message containing a tag to the cluster of the virtual document with which it is associated. Finally, it assigns new messages in the online stream to the closest clusters.

- **Sentiment140:** *Sentiment140* is a well known sentiment analysis API. The *Sentiment140* is based on the machine learning algorithms (Maximum Entropy classifier) for classifying the sentiment of Twitter messages [17]. It uses unigrams and bigrams as feature extractors.

---

[2] http://help.sentiment140.com/api.

4.2 Datasets

Three reference corpora are used in the experiments to evaluate the performances of the GLIO algorithm vis-a-vis the previously quoted reference algorithms. In this sub-section, we describe the three used reference datasets.

- **TREC 2011 Microblog Track:** In the TREC 2011, a new task called the Microblog Track is introduced to provide a benchmark for research in twitter [36]. This collection contains a sample of tweets over a period of about two weeks spanning from January 24th, 2011 to February 8th, 2011. The TREC 2011 Microblog Track collection is used to evaluate the participating real-time Twitter search systems over 50 official topics.

- **Sander:** The Sander corpus is created by [46], and is composed of 5513 tweets classified by the author into positive, negative, neutral and without importance. In addition, the corpus is labeled by topic: Apple, Google, Microsoft and Twitter. The corpus is available on the Sananalytics site[3].

- **Citeseer:** The CiteSeer corpus [14] is composed of 3312 scientific publications classified into six classes. Each publication in the dataset is described by a 0/1-valued word vector indicating the absence/presence of the corresponding word from the dictionary. The dictionary consists of 3703 single words. Given its sparseness, the data presented in the Citeseer collection are very close to the social network data. We consider every scientific publication as a user's publication, and we use this corpus to evaluate the performances of our algorithm for grouping like-minded users. Since the documents in this collection do not contain hashtags, we do not apply the SMSC algorithm on the Citeseer corpus.

Figure 4 presents the words' distributions in the reference corpora. We note, in the three corpora, a low appearing frequency (generally the frequency is lower than 100). Also, we note a big number of words having frequencies lower than 10. For example, the number of words appearing at least 10 times in the Sander corpus is 671, when the number of words appearing at most 9 times is 13776, and 9313 of words appearing once in the whole corpora.

This word distribution causes the data sparseness which characterizes data in social networks. And so, the matrix $users \times terms$ built in the text preprocessing step of our algorithm is almost full of zeros. Thus, the use of PCA in the GLIO algorithm is very important, because it allows reducing the data noise without loss of information.

4.3 Evaluation Methods

For a significant evaluation of the systems of categorization, several evaluation metrics are proposed in the literature. We present, in this section, the four evaluation measurements that we use to evaluate and compare the effectiveness of our system compared to the reference methods presented in Sect. 4.1.

---

[3] http://www.sananalytics.com/lab/twitter-sentiment/.

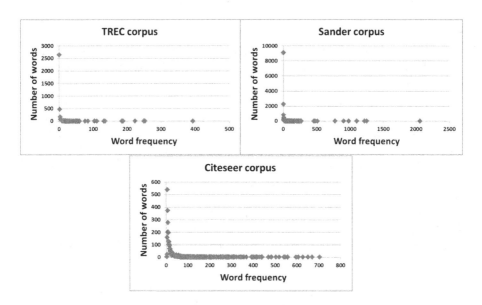

Fig. 4. Word's distribution

- **Precision:** The precision of a set of user community (cluster), determined by the system, is the proportion of correct users in this set. It is defined by:

$$Precision_i = \frac{The\ number\ of\ users\ correctly\ allocated\ to\ the\ cluster\ i}{The\ number\ of\ users\ allocated\ to\ the\ cluster\ i}$$

- **Recall:** Recall is the system capacity to retrieve all users speaking about a given subject. It is defined by:

$$Recall_i = \frac{The\ number\ of\ users\ correctly\ allocated\ to\ the\ cluster\ i}{The\ number\ of\ users\ belonging\ to\ the\ cluster\ i}$$

- **F-score:** The F-score of [44], also called total effectiveness, combines precision and recall in only one measure given by the following function:

$$F_i = \frac{2 * Recall_i * Precision_i}{Recall_i + Precision_i}$$

- **Rand-Index:** The Rand-index [43] is one of the measurements usually used for the evaluation of the clusters. It is given by the following function:

$$RI = \frac{\sum t_p + \sum t_n}{\sum t_p + \sum f_p + \sum t_n + \sum f_n}$$

Where t_p is a pair of true positive which is a pair of users (u_1, u_2) belonging to the same class C_i and is classified in the same cluster K_j. The pair true negative t_n is a pair of users (u_1, u_2) belonging to two different classes C_i and C_j and are classified in two different clusters K_i and K_j, with $i \neq j$. A pair of false positive f_p is when u_1 and u_2 belong to two different classes C_i and C_j and are classified in the same cluster K_i. The false negative f_n is a pair (u_1, u_2) belonging to the same class C_i and are classified in two different clusters K_i and K_j.

5 Results

In this section, we present the results obtained using our algorithm and the three reference methods: K-Means, LDA and SMSC. To apply the reference algorithms, we must manually provide the number of the desired clusters in the input, while, our GLIO algorithm finds this value automatically (see Sect. 3.4).

To better evaluate our algorithm for grouping like-minded users, we present in this section a qualitative and a quantitative evaluation of the grouping results. For the qualitative evaluation, we present the users' distribution into classes and clusters. In a second step, we use Recall, Precision, F-score and RI measures for the quantitative evaluation. We also evaluate the performances of the polarity detection task by comparing our obtained results with those obtained with the *sentiment140* API.

5.1 Results of the Users Grouping Task

For a complete evaluation and a better interpretation of the used systems, we compare the user distributions by cluster. To evaluate the grouping algorithms, we consider two principal criteria: entirety and homogeneity. Thus, the system must put all the users of a given class C in a same cluster K, which contains only the class C users.

Figure 5 presents the Sander corpus user distributions by cluster respectively obtained with the algorithms K-Means, LDA, SMSC and GLIO. We notice that the clusters, obtained with the K-Means algorithm, suffer from a great heterogeneity. Moreover, the users are almost arranged in only three clusters while four subjects are initially considered.

Each group obtained with LDA presents the four subjects with close proportions, whereby we can not affirm the dominating subject of each cluster. We note that the SMSC algorithm succeeded in releasing three clusters whose users belong to only one class (Cluster 2 = Twitter, Cluster 3 = Google and Cluster 4 = Microsoft). Yet, Cluster 1 contains messages belonging to the four classes. Finally, it is clear that our GLIO algorithm generates four very homogeneous clusters and each one of them contains the users of only one class (Cluster 1 = Google, Cluster 2 = Microsoft, Cluster 3 = Twitter and Cluster 4 = Apple).

Tables 2, 3 and 4 show the results of the users' grouping obtained for the three corpora (TREC, Sander and Citeseer respectively) by applying the algorithms K-Means, LDA, SMSC and GLIO. We notice that the low values are obtained with the classical approaches (K-Means and LDA). These low values highlight

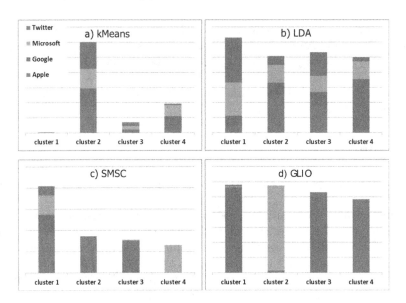

Fig. 5. Users distribution by topics and clusters

the limits of the classical approaches vis-a-vis the data sparseness problem which characterizes the social network data. We also notice that the SMSC algorithm provides good performances which exceed those obtained with LDA and K-Means for the grouping of like-minded users.

The values of recall, precision, F-score and the Rand-Index prove that the results of our GLIO algorithm exceed those of the reference algorithms. These results can be explained by the fact that the PCA in step 2 of our algorithm reduces the data noise. This latter is due to the weak density, which characterizes the corpora of the short texts exchanged in the social networks.

5.2 Results of Polarity Detection Task

In order to evaluate the results obtained using the sentiment analysis method (proposed by [6]), we need a reference classification (a labelled corpus). As the

Table 2. Results of grouping like-minded users in the TREC corpus

	TREC			
	Recall	Precision	F	RI
K-Means	0.81	0.48	0.6	0.68
LDA	0.52	0.57	0.54	0.79
SMSC	0.86	0.83	0.85	0.91
GLIO	**0.91**	**0.83**	**0.86**	**0.94**

Table 3. Results of grouping like-minded users in the Sander corpus

	Sander			
	Recall	Precision	F	RI
K-Means	0.34	0.37	0.35	0.5
LDA	0.5	0.5	0.5	0.68
SMSC	0.85	0.75	0.79	0.76
GLIO	**0.98**	**0.97**	**0.98**	**0.97**

Table 4. Results of grouping like-minded users in Citeseer corpus

	Citeseer			
	Recall	Precision	F	RI
K-Means	0.27	0.25	0.26	0.68
LDA	0.59	0.53	0.55	0.79
SMSC	-	-	-	-
GLIO	**0.65**	**0.66**	**0.65**	**0.81**

corpus ("Citeseer" documents and the "TREC" corpus tweets) does not contain information characterizing the sentiments brought in the texts, we use only the "Sanders-Twitter Sentiment" corpus. By using this corpus, we apply Sentiment140 API to compare its results with those obtained with our method. In fact, we notice that our sentiment classification method is more effective and efficient than those obtained by *Sentiment140 API*. Indeed, we obtain with our method two consistent classes (positive and negative) which have an efficiency of 76.58 % compared to reference classes. However, when we apply the *Sentiment140 API*, we find an efficiency of 66.15 %.

For a complete vision, we test the ability of our system to group users having the same opinion polarity towards a specific subject within the same group. To do so, we extract from Sanders collection corpus only the positive and negative posts. Thus, this corpus contains four subjects (Apple, Google, Microsoft and Twitter). For each subject (sub-set of posts), we classify the posts into two classes (positive and negative) using our method and the *sentiment140 API*. Therefore, we obtain eight groups of users. Each group contains users talking about the same subject and having the same polarity (positive or negative).

The Table 5 presents the grouping results of our algorithm and using the *sentiment140 API*. We note that the average of the recall, precision and F-score for the obtained clusters using the GLIO algorithm are respectively 0.81, 0.73 and 0.74. When, the averages for the clusters obtained using *sentiment140* are 0.53, 0.52 and 0.48.

Also, we obtain an average of Rand-Index equal to 0.66 and 0.62 using the GLIO and sentiment140 respectively. These results imply that our system, for

Table 5. Results of grouping like-minded users

Groups		GLIO				Sentiment140			
		Precision	Recall	F-score	RI	Precision	Recall	F-score	RI
Apple	Positive	0.47	0.67	0.55	0.55	0.43	0.51	0.47	0.87
	Negative	0.76	0.66	0.71	0.55	0.70	0.69	0.70	0.49
Google	Positive	0.85	0.92	0.88	0.85	0.80	0.64	0.70	0.79
	Negative	1	0.41	0.59	0.45	0.29	0.41	0.34	0.52
Microsoft	Positive	0.80	0.51	0.62	0.49	0.52	0.13	0.21	0.82
	Negative	0.71	0.84	0.77	0.71	0.58	0.84	0.68	0.48
Twitter	Positive	0.89	0.87	0.88	0.84	0.45	0.87	0.68	0.56
	Negative	0.98	0.94	0.96	0.85	0.5	0.05	0.1	0.47
Average		**0.81**	**0.73**	**0.74**	**0.66**	**0.53**	**0.52**	**0.48**	**0.62**

the most part, assigns the right polarity to each user, and successful groups like-minded users having similar opinions.

6   Conclusion

In this work, we presented the GLIO algorithm (Grouping Like-minded users based on Interest-centers and Opinions). Our algorithm aims to seek for the latent centers in a group of users based on their publications, and to assign each user to the nearest center. The GLIO algorithm follows five steps to group users according to their interest centers and opinions: (1) Giving a set of users' textual posts, we start by preprocessing the text to reduce the data noise characterizing posts in social networks. (2) In a second step, we retrieve latent interest centers using PCA. (3) Then, we calculate the similarities between users and the retrieved centers of interest. In the fourth step, (4) we use the K-Means algorithm to group users having similar interests. (5) Finally, we divide each group according to the user message polarity (positive or negative).

Compared with three reference algorithms of grouping like-minded users, namely K-Means, LDA and SMSC, the experimental results using three datasets (TREC 2011 Microblog Track, Sander and Citeseer) prove the effectiveness and the high quality of our algorithm results. Additionally, unlike the reference algorithms, our algorithm allows finding automatically the number of clusters. Furthermore, our algorithm used to detect the users' polarities perform better than the well-known *sentiment140*. Moreover, the proposed algorithm is flexible, as it allows replacing the clustering algorithm K-Means by another algorithm or the used similarity metric cosine by another metric as required.

In the long run, a classification part of online users may be added in order to study the network evolution. Amongst the prospects that can be considered is to add a semantic layer by integrating an ontology or Folksonomies such as Open Directory Project (ODP). Information about the links between the users can also be considered as an additional information to improve the user clustering results, and to deduce user's opinion. Finally, it is actually motivating to group users sharing the same interests and speaking different languages.

References

1. Abrouk, L., Gross-Amblard, D., Leprovost, D.: Découverte de communautés par analyse des usages. In: EGC 2010 Workshops (Workshop Web Social), pp. A5-5–A5-16 (2010)
2. Abu-Jbara, A., King, B., Diab, M.T., Radev, D.R.: Identifying opinion subgroups in arabic online discussions. In: ACL (2), pp. 829–835. The Association for Computer Linguistics (2013)
3. Adamic, L., Glance, N.: The political blogosphere and the 2004 US election: divided they blog. In: Proceedings of the 3rd International Workshop on Link Discovery, LinkKDD 2005, pp. 36–43. ACM, New York, NY, USA (2005)
4. Aggarwal, C.C., Zhai, C.: A survey of text clustering algorithms. In: Aggarwal, C.C., Zhai, C. (eds.) Mining Text Data, pp. 77–128. Springer, New York (2012)
5. Aggarwal, C.C., Zhao, Y., Yu, P.S.: On text clustering with side information. In: 2012 IEEE 28th International Conference on Data Engineering (ICDE), pp. 894–904, April 2012
6. Ameur, H., Jamoussi, S.: Dynamic construction of dictionaries for sentiment classification. In: Proceedings of the 2013 IEEE International Conference on Data Mining Workshops (ICDM 2013), Dallas, Texas, USA (2013)
7. Angelova, R.: A neighborhood-based approach for clustering of linked document collections. In: CIKM 06: Proceedings of the 15th ACM International Conference on Information and Knowledge Management, pp. 778–779. ACM Press (2006)
8. Blei, D.M., Ng, A.Y., Jordan, M.I.: Latent dirichlet allocation. J. Mach. Learn. Res. **3**, 993–1022 (2003)
9. Cattell, R.B.: The scree test for the number of factors. Multivar. Behav. Res. **1**(2), 245–276 (1966)
10. Ding, X., Liu, B.: The utility of linguistic rules in opinion mining. In: Proceedings of the 30th Annual International ACM SIGIR conference on Research and development in information retrieval, New York, NY, USA (2007)
11. Esuli, A., Sebastiani, F.: SentiWordNet: a publicly available lexical resource for opinion mining. In: Proceedings of LREC 2006, the 5th Conference on Language Resources and Evaluation, Pisa, Italy (2006)
12. Fortunato, S.: Community detection in graphs. Phys. Rep. **486**(3–5), 75–174 (2010)
13. Gemmell, J., Shepitsen, A., Mobasher, B., Burke, R.: Personalizing navigation in folksonomies using hierarchical tag clustering. In: Song, I.-Y., Eder, J., Nguyen, T.M. (eds.) DaWaK 2008. LNCS, vol. 5182, pp. 196–205. Springer, Heidelberg (2008)
14. Giles, C.L., Bollacker, K.D., Lawrence, S.: Citeseer: an automatic citation indexing system. In: International conference on digital libraries, pp. 89–98. ACM Press (1998)
15. Girvan, M., Newman, M.E.J.: Community structure in social and biological networks. Proc. Natl. Acad. Sci. **99**(12), 7821–7826 (2002)
16. Généreux, M., Santini, M.: Defi: classification de textes francais subjectifs. In: Natural Language Technology Group, University of Brighton, Brighton, United Kingdom (2007)

17. Go, A., Bhayani, R., Huang, L.: Twitter sentiment classification using distant supervision. Processing, pp. 1–6 (2009)
18. Handcock, M.S., Raftery, A.E, Tantrum, J.M.: Model-based clustering for social networks (2007)
19. Hang, C., Vibhu, O.M., Mayur, D.: Comparative experiments on sentiment classification for online product reviews. In: AAAI, Boston, Massachusetts (2006)
20. Hannachi, L., Asfari, O., Benblidia, N., Bentayeb, F., Kabachi, N., Boussaid, O.: Community extraction based on topic-driven-model for clustering users tweets. In: Zhou, S., Zhang, S., Karypis, G. (eds.) ADMA 2012. LNCS, vol. 7713, pp. 39–51. Springer, Heidelberg (2012)
21. Harb, A., Plantié, M., Dray, G., Roche, M., Trousset, F., Poncelet, P.: Web opinion mining: how to extract opinions from blogs? In: Proceedings of the 5th International Conference on Soft Computing as Transdisciplinary Science and Technology, CSTST 2008, pp. 211–217. ACM, New York, NY, USA (2008)
22. Hassan, A., Abu-Jbara, A., Radev, D.R.: Detecting subgroups in online discussions by modeling positive and negative relations among participants. In: EMNLP-CoNLL, pp. 59–70. ACL (2012)
23. Hogenboom, A., Bal, D., Frasincar, F., Bal, M., de Jong, F., Kaymak, U.: Exploiting emoticons in sentiment analysis. In: Proceedings of the 28th Annual ACM Symposium on Applied Computing, SAC 2013, pp. 703–710. ACM, New York, NY, USA (2013)
24. Jaffali, S., Jamoussi, S.: Principal component analysis neural network for textual document categorization and dimension reduction. In: 2012 6th International Conference on Sciences of Electronics, Technologies of Information and Telecommunications (SETIT), pp. 835–839 (2012)
25. Kaiser, H.F.: The application of electronic computers to factor analysis. Educ. Psychol. Measur. **20**(1), 141–151 (1960)
26. Li, X., Guo, L., Zhao, Y.E.: Tag-based social interest discovery. In: Proceedings of the 17th International Conference on World Wide Web, WWW 2008, pp. 675–684. ACM, New York, NY, USA (2008)
27. Liang, H., Xu, Y., Li, Y.: Mining users' opinions based on item folksonomy and taxonomy for personalized recommender systems. In: Fan, W., Hsu, W., Webb, G.I., Liu, B., Zhang, C., Gunopulos, D. Wu, X. (eds.) ICDM Workshops, pp. 1128–1135. IEEE Computer Society (2010)
28. Lin, Y., Sundaram, H., Chi, Y., Tatemura, J., Tseng, B.: Discovery of blog communities based on mutual awareness. In: Proceedings of the 3rd Annual Workshop on the Weblogging Ecosystem (2006)
29. Liu, J.: Comparative analysis for k-means algorithms in network community detection. In: Cai, Z., Hu, C., Kang, Z., Liu, Y. (eds.) ISICA 2010. LNCS, vol. 6382, pp. 158–169. Springer, Heidelberg (2010)
30. MacQueen, J.B.: Some methods for classification and analysis of multivariate observations. In: Le Cam, L. M., Neyman, J. (eds.) Proceedings of the Fifth Berkeley Symposium on Mathematical Statistics and Probability, vol. 1, pp. 281–297. University of California Press (1967)
31. Mahajan, M., Nimbhorkar, P., Varadarajan, K.: The planar k-means problem is NP-hard. Theor. Comput. Sci. **442**, 13–21 (2012)
32. McCallum, A., Wang, X., Corrada-Emmanuel, A.: Topic and role discovery in social networks with experiments on enron and academic email. J. Artif. Int. Res. **30**(1), 249–272 (2007)

33. McGlohon, M., Akoglu, L., Faloutsos, C.: Statistical properties of social networks. In: Aggarwal, C.C. (ed.) Social Network Data Analytics, pp. 17–42. Springer, New York (2011)
34. Newman, M.E.J., Girvan, M.: Finding and evaluating community structure in networks. Phys. Rev. E **69**, 026113 (2004)
35. Nigam, K., Hurst, M.: Towards a robust metric of polarity. In: James, G.S., Qu, Y., Wiebe, J. (eds.) Computing Attitude and Affect in Text: Theories and Applications. The Information Retrieval Series, vol. 20. Springer, Dordrecht (2006)
36. Ounis, I., Lin, J. Soboroff, I.: Overview of the trec-2011 microblog track. In: TREC (2011)
37. Pak, A., Paroubek, P.: Construction dun lexique affectif pour le franais partir de twitter. Universit de Paris-Sud, Cedex, France, TALN 2010, Juillet 2010
38. Palsetia, D., Patwary, M.M., Zhang, K., Lee, K., Moran, C., Xie, Y., Honbo, D., Agrawal, A., Liao, W., Choudhary, A.: User-interest based community extraction in social networks. In: The 6th SNA-KDD Workshop 12. ACM (2012)
39. Pearson, K.: On lines and planes of closest fit to points in space. Philos. Mag. **2**, 559–572 (1901)
40. Poirier, D.: Des textes communautaires à la recommandation. Ph.D. thesis, Université d'Orléans (2011)
41. Qi, G.-J., Aggarwal, C.C., Huang, T.: Community detection with edge content in social media networks. In: 2012 IEEE 28th International Conference on Data Engineering (ICDE), pp. 534–545, April 2012
42. Raîche, G., Walls, T.A., Magis, D., Riopel, M., Blais, J.: Non-graphical solutions for cattells scree test. Methodol. Eur. J. Res. Methods Behav. Soc. Sci. **9**(1), 23–29 (2013)
43. Rand, W.M.: Objective criteria for the evaluation of clustering methods. J. Am. Stat. Assoc. **66**(336), 846–850 (1971)
44. Van Rijsbergen, C.J.: Information Retrieval, 2nd edn. Butterworth-Heinemann, Newton (1979)
45. Sachan, M., Contractor, D., Faruquie, T. A., Subramaniam, L.V.: Using content and interactions for discovering communities in social networks. In: Proceedings of the 21st International Conference on World Wide Web, WWW 2012, pp. 331–340. ACM, New York, NY, USA (2012)
46. Sanders, N.J.: Sanders-Twitter sentiment corpus. Sanders Analytics LLC, October 2011
47. Solakidis, G.S., Vavliakis, K.N., Mitkas, P.A.: Multilingual sentiment analysis using emoticons and keywords. In: 2014 IEEE/WIC/ACM International Joint Conferences on Web Intelligence (WI) and Intelligent Agent Technologies (IAT), Warsaw, Poland, 11–14 August 2014 , vol. I, pp. 102–109 (2014)
48. Tang, L., Wang, X., Liu, H.: Community detection via heterogeneous interaction analysis. Data Min. Knowl. Discov. **25**(1), 1–33 (2012)
49. Tsur, O., Littman, A., Rappoport, A.: Efficient clustering of short messages into general domains. In: Kiciman, E., Ellison, N.B., Hogan, B., Resnick, P., Soboroff, I. (eds.) ICWSM. The AAAI Press (2013)
50. Turney, M., Peter, D., Littman, L.: Measuring praise and criticism: inference of semantic orientation from association. ACM Trans. Inf. Syst. (TOIS) **21**(4), 315–346 (2003)
51. Turney, P.: Thumbs up or thumbs down? semantic orientation applied to unsupervised classification of reviews. In: Proceedings of the Association for Computational Linguistics (ACL), Philadelphia, Pennsylvania (2002)

52. Vincze, N., Bestgen, Y.: Identification de mots germes pour la construction d'un lexique de valence au moyen d'une procédure supervisée. In: Actes de la 18e conférence sur le Traitement Automatique des Langues Naturelles, TALN 2011, Montpellier, France (2011)

53. Wang, X., Liu, H., Fan, W.: Connecting users with similar interests via tag network inference. In: The 20th ACM Conference on Information and Knowledge Management (CIKM), Glasgow, Scotland, UK (2011)

54. Wilson, T., Wiebe, J., Hwa, R.: Just how mad are you? finding strong and weak opinion clauses. In: Proceedings of AAAI, San Jose, US (2004)

55. Wilson, T., Wiebe, J., Hoffmann, P.: Recognizing contextual polarity in phrase-level sentiment analysis. In: Proceedings of the Human Language Technology Conference and the Conference on Empirical Methods in Natural Language Processing (HLT/EMNLP), Vancouver, CA (2005)

56. Yang, T., Jin, R., Chi, Y., Zhu, S.: Combining link and content for community detection: a discriminative approach. In: Proceedings of the 15th ACM SIGKDD International Conference on Knowledge Discovery and Data Mining, KDD 2009, pp. 927–936. ACM, New York, NY, USA (2009)

57. Yessenov, K.: Sentiment analysis of movie review comments (2009)

58. Zhou, D., Resnick, P., Mei, Q.: Classifying the political leaning of news articles and users from user votes. In: Adamic, L.A., Baeza-Yates, R.A., Counts, S. (eds.) ICWSM. The AAAI Press (2011)

A Preferences Based Approach for Better Comprehension of User Information Needs

Sondess Missaoui[1][(✉)] and Rim Faiz[2]

[1] LARODEC, ISG, University of Tunis, Bardo, Tunisia
sondes.missaoui@yahoo.fr
[2] LARODEC, IHEC Carthage University, Carthage Presidency, Carthage, Tunisia
Rim.Faiz@ihec.rnu.tn

Abstract. Within Mobile information retrieval research, context information provides an important basis for identifying and understanding user's information needs. Therefore search process can take advantage of contextual information to enhance the query and adapt search results to user's current context. However, the challenge is how to define the best contextual information to be integrated in search process. In this paper, our intention is to build a model that can identify which contextual dimensions strongly influence the outcome of the retrieval process and should therefore be in the user's focus. In order to achieve these objectives, we create a new query language model based on user's preferences. We extend this model in order to define a relevance measure for each contextual dimension, which allow to automatically classify each dimension. This latter is used to compute the degree of change in result lists for the same query enhanced by different dimensions. Our experiments show that our measure can analyze the real user's context of up to 12000 of dimensions (related to 4000 queries). We also show experimentally the quality of the set of contextual dimensions proposed, and the interest of the measure to understand mobile user's needs and to enhance his query.

Keywords: Mobile search · User's context · Relevance · User's preferences

1 Introduction

The heart of Mobile Information Retrieval (IR) research is how to exploit environmental and personal information at different contextual dimensions so that users receive relevant results quickly and conveniently. Regarding Mobile IR, context provides an important basis for identifying and understanding user's query. The representative information about the user's context should enable IR systems to better meet information needs. This domain is in rise according to many studies such as [5,9,16,18], and this often caused by the huge development and processing in mobile devices technologies. Those later have special

© Springer-Verlag Berlin Heidelberg 2015
N.T. Nguyen (Ed.): Transactions on CCI XVIII, LNCS 9240, pp. 67–85, 2015.
DOI: 10.1007/978-3-662-48145-5_4

characteristics which make them very advanced to sense different user's contextual dimensions. In fact, PDAs or smartphones are designed to take into account the user's situation and can ultimately allow the inference of the image or video content from the context using GPS, Bluetooth and even more sensors. On the other hand unlike the desktop systems, mobile devices usually have small screen size and tiny keypads and much more problems such as disconnected operation, low processing speed.

Beyond the special characteristics of mobile devices, people want to access information anytime and anywhere with the Smartphones they carry all the time. Within this practise, mobile information content as well as effective mobile search systems has attracted increasing attention. This suggests a latent demand for mobile IR that is able to enhance results visualization through the knowledge of the user's context and specificities. For this purpose, IR systems must take into account considerable challenges such as the users requirement (when they need information about something, they expect to get it right away), the mismatch query problem [14], the dynamic and less predictable environments, the shorter sessions compared with desktop usage. According to [9], users describe their needs in shorter query and use less number of queries by session. They usually consult only the first page of results [20].

Now faced with this evolving mobile technologies, exploiting the user context to enhance search quality became a necessity. For example, taping the query "Jobs" a user can search for a summer job, a permanent job, a job around his location (in his city), a job according to his interest (preferences) or even a statistical review about the number of job opportunities created in his country. So, for a clear distinction of the user's need we have to integrate different contextual dimensions. Often, with mobile applications, some aspects of the user's context are available, and this context can affect what sort of information is relevant to the user. The context can include a wide range of dimensions that characterize the situation of the user. But the question arises what is the impact of each contextual dimension on the quality of results? and How it can be considered as relevant to enhance the search task? What contextual dimensions reflect better the mobile user's need and lead to the appropriate search results? Our work consists in searching a new metric of dimensions impact on the search results quality.

In fact, the mobile users enter a limited number of terms in a query. This creates a big challenge to the IR systems which called "query mismatch problem" [14]. So many studies integrate different context fields to enhance the query such as [18], and especially to model the context, allowing to identify information that can be usefully exploited to improve search results such as [4,5,8,31] and [21].

In this paper, we focus our research efforts in this area that has received less attention which is the context filtering. We have brought a new approach that has addressed this issue. How to define the relevant contextual dimension accurately and rapidly? Our work has proceeded in terms of adapting the mobile context to user's preferences and identifying relevant contextual dimensions.

We propose a new approach allows to define the most relevant and influential user's context dimensions for each search situation.

In fact, our hypothesis is that an accurate and relevant contextual dimension is the one that provides an interesting improvement in the Preference query profile. Those dimensions can improve the quality of search by proposing to the user results tailored to his current situation and preferences.

The remainder of this paper is organized as follows. In Sect. 2, we give a comparison between Mobile IR and Traditional IR, then we precise the specificities of the emerging area of mobile search. In Sect. 3 we give an overview of related work which address Context-centred mobile web search. We describe in Sect. 4, the Context adaptation approach to user's preference. In Sect. 5, we discuss experiments and obtained results. Finally, Sect. 6 concludes this paper and outlines future work.

2   Mobile Information Retrieval VS Traditional Information Retrieval

Mobile Information Retrieval and Traditional Information Retrieval approaches pursued the same aim, which is to return relevant response to meet an information need from a collection of documents, but differ in their responses and the means implemented. Their fundamental difference is the nature of the information they return.

Traditional IR approaches consider that user needs are described fully by the user query. These approaches provide the same results for the same keyword queries even though these latter are submitted by different users in different contexts. Whereas, Mobile IR approaches aim to tailor search results to individual users by taking into account their intentions, preferences and dynamic physical context. These approaches aim to satisfy the information needs of a mobile user retrieving information via Smartphones.

In fact, Mobile IR is an expanding area, which take advantage from the growing diversity of Smartphone as well as the availability of a large amount of mobile media content that have been generated rapidly.

Successfully, a key distinction of Mobile search systems, makes them more advanced than their traditional counterparts, is that the user's environment is dynamic, in contrast to information searching in the desktop, where a user's environment is less likely to change. The identification of a such user's environment allows to identify information that can be usefully exploited to the aim of improving search effectiveness. Actually, by user's context we refer to the information characterizing the user's physical and social context. The contextual information may include geographic data, Time, activity, preferences of the user, etc. In fact, Smartphones have advanced features that make them able to generate such information and a more powerful understanding of the user's current context (situation). Recent advances in mobile telecommunications technology have seen the functionality of phones evolve from predominantly voice exchange to the capacity to process and gathered contextual and personal information.

Traditional IR systems that assume a stagnant user's environment are no longer suitable for such extremely mobile scenarios.

2.1 Specificities of Mobile Information Retrieval

Indeed, in mobile environment many factors bring together to present challenges issues to the IR systems. Challenges are of changing location and social context, restricted time for information access, and the need to adapt resources with concurrent physical characteristics of mobile devices. Those latter can play a main role in the paradigm of mobile search. Hence, the use of mobile devices influences not only the types of information people seek but also the ways they attempt to access. The impact of Smartphones, and mobile devices in general, in the search process is related to their specificities and characteristics. As they have more features than their computer counterparts, Smaetphones can provide efficient information where the relevant data that characterizes the user context is represented (such as location, weather, time, social networks and so on).

The availability of a context model may enhance the search process and allows to personalize it by considering several dimensions in the relevance assessment process. But at this point obvious questions arise: (1) how to make this contextual information available to an IR system? (2) how to model it? (3) and how to exploit the user context accurately in the search process?

Independently of the context model, an interesting aspect which emerges in mobile IR is that the availability of a new mobile technologies makes it necessary to process large amounts of data on small screens and in power-limited settings, which is a Content adaptation issue. Actually Mobile IR interprets both the content and the context to extract useful information and relationships, in order to proactively improve designs for mobile phones and related devices [9,16]. To this aim, two important issues that have been addressed in the literature related to Mobile IR concern Context-awareness and Content Adaptation. In this paper, we discuss the Context-awareness issue which is highly related to our research.

In the next part (cf. Section 3), we will discuss some recent research efforts, that have addressed the problem of personalizing search using user context.

2.2 Discussion

Traditional IR approaches are efficient to find relevant information with very high precision when knowing the right words to use in a search query. These approaches consider that user needs are described fully by the user query, which isn't always true in a mobile search environment.

In fact, Traditional IR approach can't provide personalized results accurate to the mobile user situation and intention when it considers the query as the main clue that specifies the user information need. Although the advances in mobile technology, Smartphones still have smaller screens, less processing power and low memory, those factors influence mobile users seeking behaviour. Since, mobile individuals become less patient and typically use one or two keywords maximum per web search and their queries shorter and more ambiguous. Those seeking habits create the query mismatch problem [14], which cannot be solved using Traditional IR approaches.

Indeed, Traditional approaches haven't the potential to enhance the user query with the user profile, interests, preferences or context that could be explicitly set by the user or gathered implicitly from the user search history. Thereby, an emerging need, of scalable algorithms that can perform well in power-limited settings, appeared. Hence, research efforts focus on efficient ways to process large amounts of data and to personalize the mobile search process.

Mobile IR is a significant task in information retrieval and when coupled with context awareness technologies they can become key tools for Mobile search applications. Context-aware computing is an interesting paradigm in which Mobile IR can take advantage of contextual techniques and algorithms to produce a search outcome in response to a user's information need, which is tailored to the specific context. In the next section we introduce and discuss Context-centred approaches for Mobile IR.

3   Context-Centred Mobile Information Retrieval

In recent years there has been an increasing research interest in the problem of contextualizing mobile search. Furthermore, with the availability of mobile devices and technologies that can detect the user's context, Context-centred Mobile IR are applied to produce a search outcome in response to a mobile user's query. In fact, exploiting the user context, become necessary to define any process aimed at tailoring user's information need and enhancing mobile search quality.

To discuss this paradigm, we present, in the following, the key notion of context. Then, we discuss some approaches proposed in the literature to exploit the user context and usefully apply it to the aim of improving search effectiveness.

3.1   Context-Awareness

Context-awareness is an expanding and vital field of research and numerous definitions of context exist because of the multidisciplinary and rich nature of the topic. There are many research efforts related to context-aware paradigm, studies from various points of view such as Schilit et al. [2], they define context as "where you are, who you are with, and what resources are nearby". This might suggest that context is more focused on the user's surrounding as opposed to his inner states. Morse et al. [1], describe context as "implicit situational information". As chosen one, we opt for the definition of Dey and Abowd [1], in which the context is: "Any information that can be used to characterize the situation of an entity. An entity is a person, place, or object that is considered relevant to the interaction between a user and an application, including the user and applications themselves. In this definition context result in relations between applications, situations, and entities."

In fact, in context-aware sensitive mobile search, approaches focuse on the aim of modelling the user's current context, and exploit it in the retrieval process. Given the specificity of Mobile devices, looking for information is facing several challenges such as recognizing the user's intention behind the query (Bouidghaghen et al. [18]), personalizing web search (Tsai et al. [5], Pitkow et al. [10], Ahn et al. [8]), enhancing the user's query with the current context to better meet the individual needs [9], understanding the nature of research practices [16], and Modelling the context ([4,19,21,29]). In fact, mobile users are often engaged in a specific task in the real world while they are moving, since their context is in a state of change, which adds further complexity. With this in mind, Context-aware approaches should use the right information about the user's current context as a means to deliver relevant content to that situation, within the purpose to build mobile search systems that consider increasing the precision of results.

In the literature several approaches have been proposed,which can be roughly categorized into three main classes:

- One dimension fits all approaches
- Approaches use a predefined set of contextual dimensions
- Context adaptation approaches

3.2 One Dimension Fits All Approaches

Regarding contextual information, some approaches which are characterized as "one dimension fits all" using one same contextual dimension to personalize all search queries regardless the query's keyword or the user's intentions behind it. These approaches consider user's context as one dimension in all search sessions. In this category, location is probably the most commonly used variable in context recognition. Several studies such as Bouidghaghen et al. [18], Welch and Cho [15], Chirita et al. [19], Vadrevu et al. [30] and Gravano et al. [13] have built models able to categorize queries according to their geographic intent.

Indeed, they identify the query sensitivity to location in order to determine whether the user's need is related to his geographical location or not. With the aim to personalize the search results using geographic dimension, many studies (Welch and Cho [15], Vadrevu et al. [30] and Gravano et al. [13]) have suggested to use the classification techniques to personalize the search results using current location. Location can be considered as an important context dimension but in this field it is not the only one, others can be taken into account. Some queries have no intent for localization (e.g. Microsoft office version, Horoscope) but they are "Time" sensitive.

With the aim of recognizing the user's intention behind the search, using a unique predefining context's dimension is not accurate. For example, when a mobile user is a passenger at the airport and he is late for check-in, the relevant information often depends on more than time or localization. It is a complex searching task. So, it needs some additional context dimensions such as activities and dates (e.g., flight number inferred from the user's personal calendar or numeric agenda). As another example let us consider a group of users are preparing for an outing with friends for the weekend. If the query "musical event" is formulated by one of them, the query evaluation should produce different contextual dimensions such as location, time and preferences.

3.3 Approaches Use a Predefined Set of Contextual Dimensions

Beyond the "one dimension fits all" approaches, Coppola et al. [20] and Castelli et al. [7] propose to use 'Here' and 'now' as the main important dimensions to get just incremental enhancements of existing retrieved results. Independently of the user's intention and preferences, this category of approaches relies on a predefined set of contextual dimensions for all queries and do not offer any context adaptation models to the specific goals of the users.

Several works (Gross and Klemke [32] and Aréchiga et al. [4]) operate including Time and Location besides others dimensions, most of them build a model of context which makes it possible to consider several new dimensions in the relevance assessment process such as demographic and social context. In Mymose system, Aréchiga et al. [23] propose a multidimensional context model, which includes four main dimensions (Spatial, Temporal, Task and Personal model). Those dimensions are supported by ontologies and thesaurus to represent the knowledge required for the system.

3.4 Context Adaptation Approaches

Actually, the quality of the personalization process is strongly related to the quality of the context's model. In fact, while all aspects of the operational mobile environment have the potential to influence the outcome search results, only a subset is actually relevant. For such reason, Some other researchers such as [3,12] try to identify the appropriate contextual information in order to better meet the specific user's demand. Kessler [3] approach proposes a cognitively plausible dissimilarity measure "DIR", to automatically identify relevant contextual information. This approach is based on the comparison of result rankings stemming from the same query posed in different contexts. Such measure aims to calculate the effects of contextual changes in the IR results. Another research effort, Stefanidis et al. [12], specify context as a set of multidimensional attributes. They identify user's preferences in terms of their ability to tailor with the context state of a query.

We find that most approaches for Context-aware information delivery within Mobile IR take into account all possible gathered contextual dimensions to narrow the search. As each dimension has a unique and different impact on the retrieved results, a big challenge appeared which is about selecting the best contextual information that may help to finally retrieve highly relevant information. With this in mind, we introduce in the next section our approach for context filtering that aims to effectively identify those contextual dimensions which are eligible to encompass the user's intention and preferences.

4   Context Adaptation to Preferences: CAP Approach

In this section, we focus our efforts on evaluating the user's context, in order to leave only a subset of relevant contextual dimensions. These, which go with the user's preferences and are able to enhance the search process.

4.1   Context Model

Context is multifaceted concept that has been studied across different research disciplines. Within Mobile IR, the context is used to overcome the limitations of the mismatch query problem [14] as well as personalization aim.

The special features of Smartphone make, location information, Time, friends contacts, and almost other user's information more accessible. All this information can be considered as context dimensions and then be used as additional features to create an efficient representation of the user's situation. In our work, the context is modelled through a finite set of special purpose attributes, called context dimensions c_i, where $c_i \in C$ and C is a set of n dimensions $\{c_1, c_2 \ldots, c_n\}$ For instance we adopt a context model that represents the user's context by only three dimensions Location, Time, Activity.

- User's physical context represented by his geographic position (Location),
- User's environmental context represented by the moment of the query submission (Time),
- And the user's organizational context (Activity).

The Fig. 1 shows the concept of user's context C with an instance of a finite set of contextual dimensions Location, Time, Activity.

User's Current Context. We consider the user's current context as the user's physical, environmental and organizational situations at the moment of search.

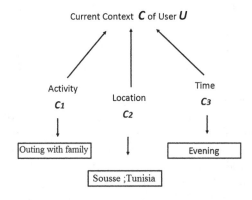

Fig. 1. An instance of a user's current context.

It can be considered as the current state at the time of the query submission. For example, when a query such as "Restaurant" is formulated by a parent, his current situation can be defined as Location: Sousse - Tunisia; Time: Evening-12/09/2012; Activity: Outing with family.

However, gathering context information is outside the scope of our research effort and our assumption is that the values of a dimension p_i can change from a search situation to another. Contextual dimension, will be defined by computing their capacity to enhance the type of retrieved documents. We evaluate their capacity to enhance the query in order to generate results with respect to the user's preferences (Preference Query Profile). In this section, we will describe our filtering model including the main features that allow to filter the user's current context and specify the most relevant contextual dimensions to narrow the search.

4.2   Preferences Model

Some recent papers have investigated language modelling approach to define the user's intention behind the query. In our work we use the language modelling approach as described in [11] to filter the context. We offer a new query language model.

1. We build a language model based preferences in our research [28]. For each user's preference, we estimate a distribution of terms associated with the user's preference. Then, we can estimate the probability that a query was issued from a given preference by sampling from the term distribution of that preference.
2. We use the query preference profile to measure the relevance of a contextual dimension.

Preferences Query Profile. According to [24]: "One way to analyze a query is to look at the type of documents it retrieves". On the basis of this rule, we infer that the best way to analyze a context dimension is to look at its effect on the query. So, its effect on the type of documents the query retrieves. Specifically, it can be accomplished by examining the top N documents of retrieval results. The context dimensions can then be ranked by the probability that they "generated" best results after being integrated in the search process. In language model approach [33] define the document likelihood of having generated the query formally as presented by the following equations:

$$P(Q \backslash D) = \prod_{w \in Q} P(w \backslash D)^{q_w} \tag{1}$$

Given a query Q and a document D, q_w is the number of times the word w occurs in query Q. According to Croft and Lafferty [35], document language models $P(w \backslash D)$, are estimated using the words in the document. We use this ranking to build a new query language model, $P(Pre \backslash Q)$, out of the top N documents. It is a new query feature in language model called "Preferences Query Profile" that helps us to define the effectiveness of the query to overcome the user's interests.

This query language model is named "the Preference profile of the query Q". In fact, a relevant retrieved result is a ranking list which meets, in a better way, the individual user needs according to their preferences. In this same spirit of thinking, we are interested in describing the personalized nature of a query. That's mean the effectiveness of the query to overcome the user's interests when it retrieving a precise topic. E.g., Searching for "Music", the mobile search system must take into account the user's preference "Jazz".

Therefore, we build a preference query profile where documents can be ranked by the probability that they have been generated depending on the user's preferences. More concretely, given a set of preferences "Pre", and a query Q, our goal is to rank the user's preferences by $P(Pre\backslash Q)$ which is initially defined as:

$$\hat{P}(Pre\backslash Q) = \sum_{D \in R} \hat{P}(Pre\backslash D) \frac{P(Q\backslash D)}{\sum_{D' \in R} P(Q\backslash D)} \tag{2}$$

Where R is the top N ranked document in a search result list and "Pre" is the name of the user's preference. It's a term that describes a user preference category from a database containing all user's interest (his profile). For example if a user is interested by "Sport" a set of terms such as (Football, Tennis, Baseball,...) are defined as "Pre".

$$P(Pre\backslash D) = \begin{cases} 1 \ if \ Pre \in Pre_D \\ 0 \quad Otherwise \end{cases} \tag{3}$$

Where Pre_D is the set of categories names of interests contained in document D (e.g. Sport, Music, News, Cinema, Horoscope ...). The profile, that describes the user's interests and preferences could be explicitly set by the user or gathered implicitly from the user search history. In our experiments, a profile is collected explicitly before starting the search session.

A very helpful step is about smoothing maximum likelihood models such as $\hat{P}(Pre\backslash Q_{in})$. We used Jelinek-Mercer process created by [6] for smoothing. We use the distribution of the initial query Q_{in}(reference-model) over preferences as a background model.

Such background smoothing is often helpful to handle potential irregularities in the collection distribution over preferences. Also, it replaces zero probability events with a very small probability. Our aim is to assign a very small likelihood of a topic where we have no explicit evidence. This reference-model is defined by:

$$\hat{P}(Pre\backslash Q_{in}) = \frac{1}{|N|} \sum_{D} \hat{P}(Pre\backslash D) \tag{4}$$

Our estimation can then be linearly interpolated with this reference model such that:

$$P'(Pre\backslash Q) = \lambda \hat{P}(Pre\backslash Q) + (1 - \lambda) \hat{P}(Pre\backslash Q_{in}) \tag{5}$$

Given λ as a smoothing parameter.

The assumption of the Preference Profile analysis is that irrelevant contextual dimension's can't improve the 'Preference Query Profile'. In fact, When we integrate an irrelevant dimension into search process, the 'Preference Query Profile' shows no variance comparing to the initial one. Given that, this contextual dimension is not important for a query and shouldn't be selected. In contrast, a relevant dimension provides a 'Preference Query Profile' with at least one peak.

Figure 2 shows an example for the effect of dimension on the Preference Language Model (preferences profile) of the query "Olympic sports". For our experiments, we choose as dimensions Time. In this example the initial query presents a flat preferences profile, while Time based profile has distinctive peaks spread over many user's preferences. Looking in depth at this graph of Fig. 2, we can see the difference between calculated preferences profile for the initial query $P(Pre \mid Q_{in}))$ and preference profile for the query enhanced by a contextual dimension c_i is "Time" $P(Pre \mid Q_{Time})$. Indeed, Time dimension makes a clear improvement for some preferences (Results, Taekwando) over the other profile. But for some others user's preferences $P(Pre \mid Q_{Time})$ is less or equal to $P(Pre \mid Q_{in}))$.

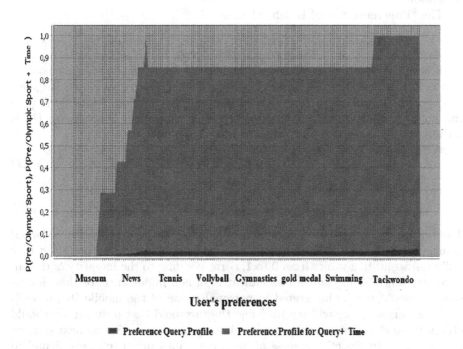

Fig. 2. Comparison between the preferences profiles of the query "Olympic sports", and enhanced queries using the contextual dimension time.

Therefore, to define the general effect of a dimension on the search process, we need a measure to specify the relevance of each dimension according to its effect on the search outcomes. In the following, we present this measure.

4.3   Preference Score Measure

Our principal objective is to adapt the user's context to his preferences automatically by filtering it. For this purpose, we need to define the most and least
influential mobile context dimensions. Indeed, there is no existing measurement
method that allows the quantification of the mobile contextual information pertinence especially using a statistical property of retrieved results list. Hence, the
task to be accomplished is to build a relevance metric measure for contextual
dimension. Our metric measure is based KL divergence [27] as an essential component to build this metric. We chose Kullback-Leibler divergence as a divergence
measure issued from the domain of probability theory. It will be introduced in
our approach to define the influence of each dimension on the search results.
The KL divergence gives us a test of similarity to the preferences background
model $P(Pre\backslash Q_{in})$. Our measure 'Preference Score' is defined basically on the
comparison of two result rankings. It allows to identify whether a mobile context
dimension enhancing the user query at his preferences profile. The "Preference
Score" is a new metric, which allows to measure the relevance degree of each
dimension.

The "Preference Score" is defined as:

$$PreferenceScore\left((P,Q)\right) = D_{kl}\left(P\left(Pre\backslash Q_p\right), P\left(Pre\backslash Q_{in}\right)\right) \tag{6}$$

Where Q is the mobile query, we denote the appearance of a dimension c_i
in a mobile context (cf. Section 4.1) as $c_i \in C$. Let $P(Pre\backslash Q_{in})$ the language
model of the initial query used as a background distribution. And $P(Pre\backslash Q_c)$
the language model of the enhanced query using contextual dimension c. D_{kl} is
a Kullback-Leibler divergence, which is initially defined as:

$$D_{kl}\left(P\left(Pre \mid Q_{c_i}\right), P\left(Pre \mid Q_{in}\right)\right) = \tag{7}$$

$$\sum_{Pre \in Pre_D} P\left(Pre \mid Q_{c_i}\right) \log \frac{P\left(Pre \mid Q_{c_i}\right)}{P\left(Pre \mid Q_{in}\right)}$$

The proposed context-based measurement model can be expressed in a formal
manner with the use of basic elements toward mathematic interpretation that
build representative values from 0 to 1, corresponding to the intensity of dimension's relevance. Being null values indicative of non importance for that dimension (it should not be integrated in personalization of the mobile IR process).
In the experiment, we will try to define the threshold that a dimension should
obtain to be classified as relevant or irrelevant information. In the next section,
we will evaluate the effectiveness of our metric measure 'Preference Score' to
classify the contextual dimensions.

5   Experimental Evaluation

Our goal is to evaluate the "Preference Score" metric to predict the type of user's
context dimension.

5.1   Dataset

For the experiments reported in this work, we used a real-world dataset which is generated through diary study. We have chosen to use a diary study for two reasons. First, the lack of a survey popular data sets for mobile search applications. Second, the diary study allows to capture data revealing the real nature of mobile user information needs. Below we describe how we managed this diary.

Diary Study and Participants. The participants (40 male, 56 female) in our study, were required to own a mobile phone and have experience searching on mobile. They include graduate students aged from 24-37, they are members of our laboratory and other ones associated with our university. The study has released in 17 weeks, we asked participants to keep a diary for this period of all their information needs and their context (Activity, Time and Location) at the search moment. We have faced a problem related to the missing data, because participants forget to record some entries. For this reason, we had relied on some experts in the field of Information Retrieval to pick manually the 4600 initial set of queries based on the signification of their terms and the presence of the required information. Those experts are graduate students in the field of Information Retrieval. Also experts have select queries which may be related to the user's environmental and physical context. After a filtering step to eliminate duplicate and navigational queries, we obtained a set of 4000 queries. Where three contextual dimensions (Time, Location and Activity) are assigned to each query to indicate the user current situation.

Search Results and Contextual Dimensions. To obtain the top N Web pages that match each query, we use a web search engine namely Google via the Google Custom Search API[1]. We considered only the first 10 retrieved results, which is reasonable for a mobile browser, because mobile users aren't likely to scroll through long lists of retrieved results.

Then, for each query in the test set we classified manually their related contextual dimensions. Each dimension is associated with a label to indicate whether it is irrelevant or relevant. The criterion to assess whether a given dimension is relevant, is based on whether the mobile user expects to see search results related to this contextual information ordered high on the results list of a search engine. E.g., for a query such as "weather" the user can express his intention to see search results related higher to location and time information. Since, these dimensions are judged relevant.

To classify contextual dimensions, we recruited 10 of our diary study participants (3 male, 7 female). They assign a pertinence degree to each dimension according to their related queries. These steps left us, in our sample test queries, with 10 % noise dimensions, 24 % irrelevant dimensions and 65,6 % relevant dimensions.

---

[1] https://developers.google.com/custom-search/.

5.2   Classification Performance of Our Metric Measure

Our experimental design allows us to evaluate the effectiveness of our technique to identify user's relevant contextual fields. For this purpose, we propose an evaluation methodology of obtaining results using manually labelled contextual dimensions. In fact, a contextual dimension's class is correct only if it matches the labelled results. Using the "Preference Score" as a classification feature, we build a context intent classifier.

In order to compute the performance of the classifiers in predicting the dimensions classes, we use standard precision, recall and F-measure measures. We use also classifiers implemented as part of the Weka[2] software. We test the effectiveness of several supervised individual classifiers (Decision trees, Naive Bayes, SVM, and a Rule-Based Classifier) in classifying contextual fields using "Preference Score" as classification feature.

5.3   Results and Discussion

Analysis of Preference Score Measure. At this level we analyse the "Preference Score" distribution for each category of contextual dimensions. Fig. 3 shows the distribution of our measure over different values of Location dimension for different queries. In this figure we notice that there is remarkable drops and peaks in the value of "Preference Score".

Moreover, this distribution of this measure for location dimensions, presented in Fig. 3, has a clear variation with multiple values which clearly support our assumption here. Indeed, the relevance of a contextual dimension is independent of his type or value but it depends on the query and the intention of mobile user behind such query. Hence, "Preference Score" measure hasn't a uniform distribution for those dimensions. It still depends on the user's query. We can conclude that the measure based on the language model approach succeeds to measure the sensitivity of user's query to each contextual dimension.

Table 1 presents the two lowest and two highest values for each dimension class, obtained from our sample test queries and their user's context. Those values allow to confirm a possible correlation between dimension intent class and "Preference Score" feature. Hence, we define a threshold value for each dimension class. And in the following, we will evaluate the effectiveness of thus thresholds to classify those dimensions.

Effectiveness of Contextual Dimension Classification. Our goal in this evaluation is to assess the effectiveness of our classification attribute "Preference Score" to identify the type of dimension of classes: relevant, irrelevant, and noise.

As discussed above, we tested different types of classifiers and Table 2 presents the values of the evaluation metrics obtained by each classifier. In fact, all the classifiers were able to distinguish between the three contextual dimension classes. Fmeasures, Precision and Recall ranging from 96 % to 99 %. But "SVM"

---

[2] http://www.cs.waikato.ac.nz/ml/weka/.

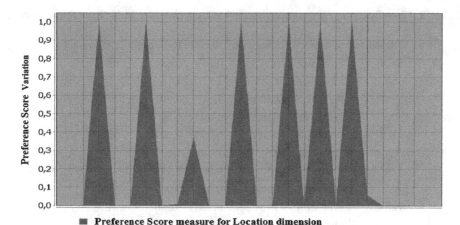

Fig. 3. Distribution of "Preference Score" measure for geographic dimension (Location).

classifier achieves the highest accuracy with 99 % for the F-measure. This experiment implies the effectiveness of our approach to accurately distinguish the three types of user's current context levels. It especially allows to correctly identify the irrelevant contextual information with an evaluation measure over 1. When relevant and noise, achieving over 97 % classification accuracy.

In a second experiment, we evaluated the classification effectiveness of our approach comparatively to the DIR approach developed by Kessler [3]. By using the DIR measure, contextual information is only classified as relevant or irrelevant. It enables distinguishing between irrelevant and relevant context using a threshold value δ. Whence, we compared the two approaches only on this basis.

We implemented the DIR approach using the SVM classifier which achieves one of the best classification performance using one simple rule: analysing the

Table 1. The lowest and highest values for each contextual dimension classes relevant and irrelevant.

Query	Dimensions	Relevance score	Class
"house for sale"	Activity: At home	0.432	Relevant
"check in airport"	Time: 12/09/2012	0.550	Relevant
"house for sale"	Location: Tunisia	0.985	Relevant
"check in airport"	Activity: Walking	0.999	Relevant
"outdoor tiki hut bars"	Time: 30/08/2012	0.147	Irrelevant
"eye chart"	Activity: Working	0.169	Irrelevant
"weather"	Activity: Walking	0.199	Irrelevant
"new bus federation"	Activity: Study	0.279	Irrelevant

Table 2. Classification performance obtained using a classifier with "Preference Score".

Classifier	Class	Precision	Recall	F-measure	Accuracy
SVM	relevant	0.978	0.989	0.981	99 %
	irrelevant	1	1	1	
	noise	0.981	1	0.991	
	average	0.991	0.99	0.99	
JRIP rules	relevant	0.911	0.953	0.924	96.3 %
	irrelevant	1	1	1	
	noise	1	0.964	0.926	
	average	0.965	0.962	0.962	
Bayes	relevant	1	0.933	0.966	97 %
	irrelevant	1	1	1	
	noise	0.946	1	0.972	
	average	0.973	0.971	0.971	
J48	relevant	1	0.933	0.966	97 %
	rrelevant	1	1	1	
	noise	0.946	1	0.972	
	average	0.973	0.971	0.971	

individual results in two rankings for the same query expanded by different contextual dimensions. Intended or relevant contextual information must have an impact that goes beyond a threshold value. Hence, we should obtain a high value of DIR measure to classify a context as relevant. Table 3 presents the precision, recall, F-measure and accuracy achieved by the SVM classifier according to the both approaches. The result of comparison show that, our approach gives higher classification performance than DIR approach with an improvement of 1 % at accuracy. This improvement is mainly over Relevant context dimensions with 1.3 % at Recall.

Table 3. Classification performance on relevant and irrelevant dimensions: comparison between CAP approach and DIR measure approach.

Approach	DIR approach			CAP approach					
Class	Relevant	Irrelevant	Average	Relevant	Impro	Irrelevant	Impro	Avrege	Impro
Precision	1	0.834	**0.917**	1	**0 %**	0.942	**0.11 %**	**0.991**	**0.86 %**
Recall	0.874	1	**0.937**	0.934	**0.06 %**	1	**0 %**	**0.967**	**0.03 %**
F-measure	0.905	0.893	**0.899**	0.981	**1.3 %**	0.992	**0.9 %**	**0.99**	**1 %**
Accuracy	**92 %**			**95 %**					**3 %**

6   Conclusion

We proposed in this paper a new approach for mobile context adaptation to the user's preferences. It evaluates the relevance of contextual dimensions using different features. This approach is based a new metric "Preference Score", that allows to classify the contextual dimensions according to their relevance to enhance the search results. Our experimental evaluation shows the classification performance of our metric measure compared to a cognitively plausible dissimilarity measure namely DIR. For future work, we plan to exploit our proposed approach to personalize mobile Web search. We will customize the search results for queries by considering the determined user's contextual dimension classified as relevant.

References

1. Abowd, G.D., Dey, A.K.: Towards a better understanding of context and context-awareness. In: Gellersen, H.-W. (ed.) HUC 1999. LNCS, vol. 1707, p. 304. Springer, Heidelberg (1999)
2. Schilit, B., Adams, N., Want, R.: Context-aware computing applications. In: Proceedings of the Workshop on Mobile Computing Systems and Applications, pp. 85–90. IEEE Computer Society, Santa Cruz, CA (1994)
3. Kessler, C.: What is the difference? a cognitive dissimilarity measure for information retrieval result sets. J. Knowl. Inf. Syst. **30**(2), 319–340 (2012)
4. Aréchiga, D., Vegas, J., Redondo, P.F.: Ontology supported personalized search for mobile devices. In: Proceedings Third International Workshop on Ontology, Conceptualization and Epistemology for Information Systems, Software Engineering and Service Science, ONTOSE. Springer LNCS, Amsterdam, pp. 1–12, 8–12 June 2009
5. Tsai, F.S., Etoh, M., Xie, X., Lee, W.C., Yang, Q.: Introduction to mobile information retrieval. J. IEEE Intell. Syst. **25**(1), 11–15 (2010)
6. Jelinek, F., Mercer, R.L.: Interpolated estimation of markov source parameters from sparse data. In: Proceedings of the Workshop on Pattern Recognition in Practice, pp. 381–397. North-Holland, Amsterdam, The Netherlands, May 1980
7. Castelli, G., Mamei, M., Rosi, A.: The whereabouts diary. In: Hightower, J., Schiele, B., Strang, T. (eds.) LoCA 2007. LNCS, vol. 4718, pp. 175–192. Springer, Heidelberg (2007)
8. Ahn, J., Brusilovsky, J.P., He, D., Grady, J., Li, Q.: Personalized web exploration with task modles. In: Proceedings of WWW 2008 the 17th International Conference on World Wide Web, Beijing, China, pp. 1–10, 21–25 April 2008
9. Hollan, J.D., Sohn, T., Li, K.A., Griswold, W.G.: A diary study of mobile information needs. In: Proceedings of SIGCHI Conference on Human Factors in Computing Systems, pp. 433–442. ACM, Florence, Italy, 5–10 April 2008
10. Pitkow, J., Schutze, H., Cass, T., Cooley, R., Turnbull, D., Edmonds, A., Adar, E., Breuel, T.: Personalized search. Commun. ACM J. **45**(9), 50–55 (2002)
11. Ponte, J.M., Croft, W.B.: A language modeling approach to information retrieval. In: Proceedings of the 21st International ACM SIGIR Conference on Research and Development in Information Retrieval, pp. 275–281. ACM, Melbourne, Australia, August 1998

12. Stefanidis, K., Pitoura, E., Vassiliadis, P.: Adding context to preferences. In: Proceedings of ICDE IEEE 23rd International Conference on Data Engineering, Istanbul, Turkey, pp. 846–855, 15–20 April 2007
13. Gravano, L., Hatzivassiloglou, V., Lichtenstein, R.: Categorizing web queries according to geographical locality. In: Proceedings of CIKM 2003 the Twelfth International Conference on Information and Knowledge Management, pp. 325–333. ACM, New Orleans, Louisiana, USA, 2–8 November 2003
14. Arias, M., Cantera, de la Fuente, P., Llamas, C., Vegas, J.: Knowledge-based thesaurus recommender system in mobile web search, In: Proceedings of CERI 1st Spanish Conference on Information Retrieval, Madrid, Spain, 15–16 June 2010
15. Welch, M., Cho, J.: Automatically identifying localizable queries. In: Proceedings of 31st Annual International ACM SIGIR Conference on Research and Development in Information Retrieval, pp. 1185–1186. ACM, Singapore, 20–24 July 2008
16. Kamvar, M., Baluja, S.: A large scale study of wireless search behavior : Google mobile search. In: Proceedings of the SIGCHI 2006 the SIGCHI Conference on Human Factors in Computing Systems, pp. 701–709. ACM, Montreal, Quebec, Canada, 22–27 April 2006
17. Matsuda, N., Hirashima, T., Nomoto, T., Taki, H., Toyoda, J.I.: Context-sensitive filtering for the Web. J. Web Intell. Agent Syst. 1(3), 249–257
18. Bouidghaghen, O., Tamine, L., Boughanem, M.: Context-aware user's interests for personalizing mobile search. In: Proceedings 12th IEEE International Conference on Mobile Data Management, pp. 129–134. IEEE Computer Society, Sweden, 6–9 June 2011
19. Chirita, P., Firan, C., Nejdl, W.: Summarizing local context to personalize global web search. In: Proceedings of CIKM International Conference on Information and Knowledge Management, pp. 287–296. ACM, Arlington, Virginia, USA, 6–11 November 2006
20. Coppola, P., Della Mea, V., Di Gaspero, L., Menegon, D., Mischis, D., Mizzaro, S., Scagnetto, I., Vassena, L.: The context-aware browser. J. IEEE Intell. Syst. 25(1), 38–47 (2010)
21. Ingwersen, P., Jarvelin, K.: The Turn: Integration of Information Seeking and Retrieval in Context, vol. 18, p. 448. Springer, New York (2005)
22. Roman, P.E., Dell, R.F., Velasquez, J.D., Heufeman, P.L.: Identifying user sessions from web server logs with integer programming. J. Intell. Data Anal. 18(1), 43–61
23. McParlane, P.J., Moshfeghi, Y., Jose, J.M.: On contextual photo tag recommendation. In: Proceedings of SIGIR 2013 the 36th International ACM SIGIR Conference on Research and Development in Information Retrieval, pp. 965–968. ACM, Dublin, Ireland, 28 July –01 August 2013
24. Jones, R.: Temporal profiles of queries. J. ACM Trans. Inf. Syst. TOIS 25(3), 14 (2007)
25. De Virgilio, R., Torlone, R.: Modeling heterogeneous context information in adaptive web based applications. In: Proceedings of ICWE 6th International Conference on Web Engineering, pp. 56–63. ACM, Palo Alto, California, USA, 11–14 July 2006
26. Cronen-Townsend, S., Zhou, Y., Croft, W.B.: Predicting query performance. In: Proceedings the 25th Annual International ACM SIGIR Conference on Research and Development in Information Retrieval, pp. 299–306. ACM, Tampere, Finland, 11–15 August 2002
27. Eguchi, S., Copas, J.: Interpreting Kullback Leibler divergence with the Neyman-Pearson lemma. J. Multivariate Anal. 97, 2034–2040 (2006)

28. Missaoui, S., Faiz, R.: A new preference based model for relevant dimension identification in contextual mobile search. In: Noubir, G., Raynal, M. (eds.) Networked Systems. LNCS, vol. 8593, pp. 215–229. Springer, Heidelberg (2014)

29. Poslad, S., Laamanen, H., Malaka, R., Nick, A., Buckle, P., Zipf, A.: Crumpet, Creation of user-friendly mobile services personalised for tourism. In: Proceedings of the Second International Conference on 3G Mobile Communication Technologies, Conference Publication No. 477, pp. 28–32. IEEE Computer Society, London, 26–28 March 2001

30. Vadrevu, S., Zhang, Y., Tseng, B., Sun, G., Li, X.: Identifying regional sensitive queries in web search. In: Proceedings of WWW 2008 the 17th International Conference on World Wide Web, pp. 1185–1186, Beijing, China, 21–25 April 2008

31. Yau, S., Liu, H., Huang, D., Yao, Y.: Situation-aware personalized Information retrieval for mobile internet. In: Proceedings of COMPSAC 27th Annual International Computer Software and Applications Conference, pp. 639–644. IEEE Computer Society, Dallas, TX, USA, 3–6 November 2003

32. Gross, T., Klemke, R.: Context modelling for information retrieval: requirements and approaches. IADIS Int. J. WWW/Internet $6(1)$, 29–42 (2003)

33. Lavrenko, V., Croft, W.B.: Relevance-based language models. In: Proceedings of SIGIR 2001 the 24th Annual International ACM SIGIR Conference on Research and Development in Information Retrieval, pp. 120–127. ACM, New Orleans, Louisiana, USA, 9–13 September 2001

34. Varma, V., Sriharsha, N., Pingali, P.: Personalized web search engine for mobile devices. In: Proceedings of IIIA International Workshop on Intelligent Information Access, Marina Congress Center, Helsinki, Finland, 6–8 July 2006

35. Croft, W.B., Lafferty, J. (eds.): Language Modeling for Information Retrieval. Kluwer Academic Publishers, Dordrecht (2003)

Performance Evaluation of the Customer Relationship Management Agent's in a Cognitive Integrated Management Support System

Marcin Hernes[(✉)]

Wrocław University of Economics, Wrocław, Poland
marcin.hernes@ue.wroc.pl

Abstract. The biggest problem currently, turns out to be the processing of unstructured knowledge in integrated management support systems. Note that knowledge contained in these systems is normally structuralized and the systems employ various methods for processing structuralized knowledge. However, in contemporary companies, unstructured knowledge is essential, mainly due to the possibility to obtain better flexibility and competitiveness of the organization. The users' opinions about products can serve as example. Therefore, unstructured knowledge supports structuralized knowledge to a high degree. This paper presents the issues related to the sentiment analysis of customers' opinions performed by Customer Relationship Management agent running in multi-agent Cognitive Integrated Management Information System. This system is an application of computational collective intelligence and allows for supporting the management processes related with all the domain of enterprise's functioning. The agents are based on the Learning Intelligent Distribution Agent cognitive architecture, described shortly in the first part of the paper. Next, the logical architecture of Cognitive Integrated Management Information System are described. The main part of article presents issues related to functionality and implementation of Customer Relationship Management agent aims to sentiment analysis. The results of research experiment, aims to performance evaluation, are presented at the last part of article.

Keywords: Integrated management information systems · Enterprise resource planning · Cognitive agents · Decision making · Sentiment analysis · Computational collective intelligence applications

1 Introduction

Integrated Management Information Systems (IMIS), equated also with ERP (Enterprise Resource Planning) systems, play an essential role nowadays in the operation of companies, being one of the most important solutions that allow to gain competitive advantage. Note that in the age of information, the entire economy is based on information and knowledge, therefore companies must employ systems which allow to collect, process and send large volumes of information as well as draw conclusions from the information, i.e. create knowledge of an organization. Contemporary IMIS

© Springer-Verlag Berlin Heidelberg 2015
N.T. Nguyen (Ed.): Transactions on CCI XVIII, LNCS 9240, pp. 86–104, 2015.
DOI: 10.1007/978-3-662-48145-5_5

exemplify such features, they are already commonly used by the companies and are characterized by full integration both at the system/application level and the business process level. Note, however, that the properties of contemporary IMIS are becoming more and more inadequate. Apart from collecting and analyzing data and generating knowledge, the system should also be able to understand the significance of phenomena occurring around the organization. It is becoming more and more necessary to make decisions based not only on knowledge but also on experience, thus far regarded as purely human domain [4]. In order to accomplish tasks set by IMIS, a multi-agent system can be used consist of several cognitive agents. They not only enable quick access to information and quick search for the required information, its analysis and conclusions, but also, besides being responsive to environment stimuli, they have cognitive abilities that allow them to learn from empiric experience gained through immediate interaction with their environments [14], which consequently allows a number of decision versions to be automatically generated and to make and execute decisions. The cognitive information processing is detailed presented at [28].

One of the sub-systems of IMIS is CRM sub-system supporting ensuring the best company-customer relations and collecting information in the customers' preferences in terms of product purchase in order to increase sales. The CRM sub-system agents' goal is, among other, analysis a sentiment (polarity) customer's opinion about products (i.e. in order to answer the questions: is an opinion positive or negative? or is the feature of product negative or positive?). These opinion can be placed, for example, on online shops, internet forums or portals. On the basis of results of such analysis, with a certain probability, can be predict how it will be developed the sale of the product. First and foremost, however, it can be determine what features of the product to be analyzed are most desired by customers.

The purpose of this paper is to present the issues related to, functionality, implementation and performance evaluation of CRM agent's. The goal of this agent is to analysis a sentiment of customer's opinions. The agent running in the scalable and open IMIS based on Learning Intelligent Distribution Agent (LIDA) architecture. The system is named CIMIS (Cognitive Integrated Management Information System) and it is an application of computational collective intelligence [18, 19] methods in form of multi-agent system.

The first part of article presents the state on the art in the field. Next the structure and functioning of the LIDA agent and the logical architecture of CIMIS is presented. The implementation and functionality of sentiment analysis agent and research experiment aims to performance evaluation are presented at the last part of article.

2 Related Works

There is no unequivocal definition that specifies what sub-systems form an IMIS and, in addition, according the surveys presented at work [17], there are only 326 publications, during the period 1997–2010, relates with enterprise management systems.

The conception of such class of systems is presented in the paper [21]. It is proposed to build a multi-agent system that could integrate information systems to support the functioning of the different business (decision) areas of enterprise, such, as:

- Enterprise Resource Planning (ERP), understand as a manufacturing management,
- Customer Relationship Management (CRM),
- Supply Chain Management (SCM),
- e-Business,
- electronic market,
- business intelligence,
- e-banking.

However, the analysis of subject literature and practical solutions [1, 2, 5, 16, 21, 29, 35] allows to systematize the architecture of the system, concluding that it is composed of the following sub-systems:

- fixed assets,
- logistics,
- manufacturing management,
- human resources management,
- financial and accounting,
- controlling,
- CRM,
- business intelligence.

Taking into consideration the cognitive agents' architectures, in the study [4] considering the taxonomy of cognitive agent architectures with respect to memory organization and learning mechanism, three main groups of the architectures were distinguished:

1. Symbolic architectures which use declarative knowledge included in relations recorded at the symbolic level, focusing on the use of this knowledge to solve problems. This group of architectures includes, among others: State, Operator And Result [15], CopyCat [12], Non-Axiomatic Reasoning System [32].
2. Emergent architectures using signal flows through the network of numerous, mutually interacting elements, in which emergent conditions occur, possible to be interpreted in a symbolic way. This group of architectures includes, among others: Cortronics [8], Brain-Emulating Cognition and Control Architecture [22].
3. Hybrid architectures which are the combinations of the symbolic and emergent approach, combined in various ways. This group of architectures includes, among others: CogPrime [7], Cognitive Agents Architecture [11], The Learning Intelligent Distribution Agent (LIDA) [6].

The problems of the sentiment (polarity) of opinions analysis are related to analysis of text documents. There are two categories of this process in the literature of the subject [27]:

- Deep Text Processing, that is a linguistic analysis of all possible interpretations and grammatical relationships occurring in natural text. The full analysis can be very complex. Moreover in many cases, obtained in this way, the information may not be necessary. For this reason, more and more often there is a tendency to carry out only a partial analysis of the text, which may be much less time-consuming and is a compromise between precision and performance.

- Shallow Text Processing, is defined as the analysis of the text to which the effect is incomplete in relation to the deep analysis of the text. The usual limitation lies in the identification of non-recursively structures or of limited recursion level, which can be diagnosed with a large degree of certainty. The structure requiring a complex analysis of the many possible solutions are overlooked or analysed in part. The analysis is addressed mainly at recognizing proper names, noun phrases, verb groups without resolving their internal structure and function in a sentence. In addition, recognized there are some main parts of sentences, for example. the judgment or the judgment group.

There are several different methods within these categories. For information retrieval the Boolean Logic Model (BML) or ranked-output systems are used [28]. A BML query consists of words or phrases concatenated with logical operators, such as AND, OR, and NOT. As a result, the set of documents is divided into two sub-sets: the first sub-set consist of documents matched the query and the second sub-set consist of documents mismatched to the query. A ranking system, using vector algebra, assesses the probability of the content of documents to the content of the query, and on this basis the ranking of the found documents is created. In this approach, for example, the Vector Space Model, Probabilistic Model or Interface Network Model are used [24].

One of the methods used in the shallow analysis of text documents is machine learning [23]. Under this method, are used, among other things, the naive Bayesian classifiers and support vector machines [13]. In these approaches is made an analysis of the prevalence of individual words (terms) in the documents concerned. For example, at work [33] with the use of support vector machines was determined, what the product attribute is considered part of the text, whereas at work [34] polarity the opinion was expressed during the passage of the text.

Another method of both deep and shallow analysis of text documents is the use of rules, on the basis of which identification (annotation) pieces of text, for a specific topic, is performed. Such rules are based on templates, taking into account the relationship between words and semantic classes of words [26]. The basis for the generation of rules can be automatic or manual analysis of the annotated corpus [20]. An analysis of documents by using rules is to assign identified the importance of text fragments in accordance with the principles enshrined in the rule, which in certain cases can be thought of as assigning the document to the category. The analysis of text documents with the use of the rules has been used, inter alia, to the extraction of spatial relationship [37], identify the requirements for the it projects, expressed on Internet forums [31] or extraction of information from real estate ads [20]. The paper [30] instead, concerns the use of semantic roles in the process text documents analysis.

In this paper, for text document analysis, it was decided to use the architecture of cognitive agent program The Learning Intelligent Distribution Agent (LIDA) developed by Cognitive Computing Research Group. This architecture is used in order to build a CIMIS prototype.

3 The LIDA Cognitive Architecture

The realization of the CIMIS is based on the LIDA cognitive agent architecture [6], which is of emergent-symbolic nature, owing to which the processing of both structured and unstructured knowledge is possible. In addition, the Cognitive Computing Research Group established by S. Franklin, elaborated in 2011 the framework (in Java language) significantly facilitating the implementation of the cognitive agent. It should also be emphasized that the whole framework code is open, i.e. the developer has access to the definitions of all methods, as opposed to, for instance, Cougaar architecture framework software, in which the agent's software code constitutes the so-called "blackbox". The LIDA cognitive agent's architecture consist of the following modules [3, 6]:

- sensory memory,
- perceptual memory,
- workspace,
- episodic memory,
- declarative memory,
- attentional codelets,
- global workspace,
- action selection,
- sensory-motor memory.

In the LIDA architecture it was adopted that the majority of basic operations are performed by the so-called codelets, namely specialized, mobile programs processing information in the model of global workspace [6]. The functioning of the cognitive agent is performed within the framework of the cognitive cycle and it is divided into three phases: the understanding phase, the consciousness phase and the selection of actions and learning phase. At the beginning of the understanding phase the stimuli received from the environment activate the codelets of the low level features in the sensory memory [3]. The outlets of these codelets activate the perceptual memory, where high level feature codelets supply more abstract things such as objects, categories, actions or events. The perception results are transferred to workspace and on the basis of episodic and declarative memory local links are created and then, with the use of the occurrences of perceptual memory, a current situational model is generated; it other words the agent understands what phenomena are occurring in the environment of the organization. The consciousness phase starts with forming of the coalition of the most significant elements of the situational model, which then compete for attention so the place in the workspace, by using attentional codelets. The contents of the workspace module is then transferred to the global workspace, simultaneously initializing the phase of action selection. At this phase possible action schemes are taken from procedural memory and sent to the action selection module, where there compete for the selection in a given cycle. The selected actions activate sensory-motor memory for the purpose of creating an appropriate algorithm of their performance, which is the final stage of the cognitive cycle [6]. The cognitive cycle is repeated with the frequency of 5–10 times per second.

Parallely with the previous actions the agent's learning is performed, which is divided into perceptual learning concerning the recognition of new objects, categories, relations; episodic learning which means remembering specific events: what, where, when, occurring in the workspace and thus available in the awareness; procedural learning, namely learning new actions and action sequences needed for solving the problems set; conscious learning relates to learning new, conscious behaviours or strengthening the existing conscious behaviours, which occurs when a given element of the situational model is often in the module of current awareness. The agent's learning may be performed as learning with or without a teacher.

It is worth emphasizing that LIDA agent have the ability of grounding the symbols, namely assign relevant real world objects to specific symbols of the natural language. This is necessary to correctly process unstructured knowledge saved mainly by means of the natural language and thus, for instance, the clients' opinions on products.

The next part of article describes the architecture of CIMIS based on the LIDA cognitive agents.

4   The CIMIS System Architecture

The CIMIS system is dedicated mainly for the middle and large manufacturing enterprises operating on the Polish market (because the user language, at the moment, is a Polish language). The CIMIS consist of following sub-systems: fixed assets, logistics, manufacturing management, human resources management, financial and accounting, controlling, CRM, business intelligence.

The fixed assets sub-system includes support for the realization of processes related to fixed asset and involved their depreciation.

The logistics sub-system has all the main features supporting the employees of logistics department in their effective work [10]. The logistics sub-system enables maintaining optimal stock to meet the needs of production department.

The manufacturing management sub-system support a processes related to a manufacturing execution. It include functions from the scope of the technical preparation of production capacity, production planning, material consumption planning, planning and execution of a manufacturing tasks, manufacturing control, visualization, monitoring and archiving.

The human resources management sub-system supports realization of such processes, as the employees of the company data and contract registering, recording of working time, wage calculation, creating the tax and social security declaration.

The financial-accounting sub-system supports registering, to the full extent, economic events, also provides important, from the point of view of business management, information, concerning, inter alia, payment capacity, revenues, costs, financial result.

Controlling sub-system is automatically processing data related to profit and loss account in cooperation with accounting sub-system. The controlling sub-system consist of both a strategic and operational controlling.

The CRM sub-system is engaged in matters connected with ensuring the best company-customer relations and collecting information in the customers' preferences in terms of product purchase in order to increase sales. The enterprise's environment monitoring is also realized by this sub-system.

The purpose of business intelligence sub-system is to enable easy and safe access to information in a company, operation of its analysis and distribution of reports within the company and among its business partners, which in turn enables quick and flexible decision making. In the context of, most of all, the business intelligence sub-system, but other sub-systems as well, the CIMIS makes cognitive visualization [36] features available, meaning it enables a visualization of multi-dimensional data in one picture that allows to find the source of a problem in a short time and contributes to creating new knowledge about an object or problem [25].

Considering the fact, that the structure of CIMIS uses cognitive agents, Fig. 1 presents the logical architecture of the system.

The system assumes that all agents are at 'not-taught' status in the initial phase. They can be initially grouped according company's needs for sub-systems. For instance, one group of agents is assigned to the logistics sub-system, another one is assigned to the manufacturing management sub-system and yet another one to financial and accounting sub-system. Within the groups, the agents can be initially 'taught' by the company that implements the system. Next stages of learning for both grouped and ungrouped agents are done by the company staff. Agents can also learn without teacher through analyzing the results of their decisions.

The agents of all sub-systems cooperate themselves in order to better business processes realization. For example, the enterprise's environment monitoring results performed by CRM sub-system agent are using by the other agents.

The main operating purpose of the Supervisor agent is to monitor the proper operation of other agents, mainly in the field of detection and solving conflicts of knowledge and experience. The agent analyzes, in close-to-real time, the structures of knowledge and experience of all agents. Whenever a conflict occurs, it employs a solution algorithm based on a method that uses consensus theory [9, 21], and the result of the agent's actions is accepted by the system as current state of knowledge and experience.

Note that all CIMIS sub-systems are connected by a single, coherent stream of information and knowledge available online to the management, because nowadays attention is paid to functional complexity, managing all fields of operation in a company, proper flow of information and knowledge among sub-systems as well as the ability to perform a variety of analyses and to create reports for management. The implementation of this solution is realized as follow:

1. Communication between modules of agents architecture was ensured by using LIDA framework's codelets,
2. Communication between agents is based on Java Message Service (JMS) technology. The representation of information and knowledge (generated in result of agents' operating) in form of XML format document, was adopted (the JMS messaging is at the text type). The communication is realized in publish/subscribe messaging domains – it guarantee, that information or knowledge generated by one of agents is immediately available for the other agents. The asynchronous message consumption is used.

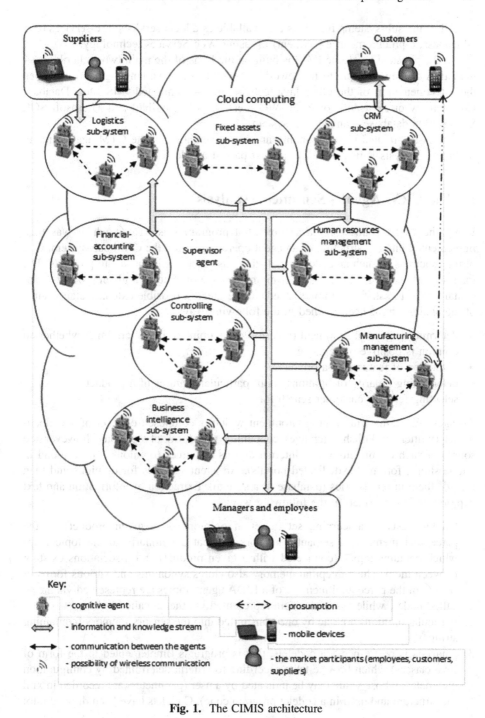

Fig. 1. The CIMIS architecture

All of the sub-systems functions are available as a local services or e-services (e.g. e-business, e-procurement, e-payment) by using Web Services technology.

At the physical level, the IMIS is built on the basis of the main two technologies – the LIDA framework (due to framework is developed at Java language and it is open the implementation of the other Java technologies – mentioned JMS, Java Database Connectivity or Java API for XML Web Services - is possible) and Microsoft SQL Server 2008 database management system.

The issues related to functionality and implementation of the CTM agent aims to sentiment analysis is presented in the next part of article.

5   The CRM Agent - Sentiment Analysis

Using the LIDA framework, a section of a program was written which was then implemented into the structure of an agent operating within the CRM sub-system. One of the functions of such an agent is analyzing customers' opinions about products. The main aim of such an analysis is determining a range of features a given products shall contain so as to satisfy customers' needs to the greatest possible extent. Detailed aims of the analysis have been defined in the following way:

- determining general sentiment (polarity) of an opinion, i.e. determining whether an opinion is positive or negative,
- extraction of features of a product,
- determining polarity of opinions about particular features of a product,
- selecting features customer search for.

An agent functions within an environment which is composed of a set of text documents (written in Polish language) containing this form of opinions, however the source of such documents is the Internet and its resources (opinions can be found in online shops, forums, or dedicated portals). An agent searches for opinions and then records them in repositories (database of a system). Extraction of information and text exploration is performed in the following way:

1. On the basis of a learning set (a set of opinions on a given product), in the perceptual memory, a semantic network is created (similarly to the topic map) which contains topics (connected with a given product) and associations existing between them. The perceptual memory also stores synonyms and various forms of words. In the perceptual memory of a LIDA agent, topics are represented via the so called nodes, while association are represented via the so called links.
2. Particular opinions are one by one transmitted into the sensory memory (containing strings).
3. Analysis is performed via codelets, that is programs (implemented in the form of Java classes) which browse texts according to criteria determined by configuration parameters whose value may be indicated by a user (parameters are recorded in xml file structure and used in a codelet program code). Codelets have been divided into:

   - codelets determining polarity of opinions,
   - codelets of extraction of features and opinions about features.

```
<task name="positive_opinion">
<tasktype> CodeletObjectDetector </tasktype>
<param name="object" type="string">good</param>
<param name="noobject" type="string">no</param>
<param name="distance" type="int">1</param>
<param name="node"
type="string">positive_opinion</param>
</task>

<task name="negative opinion">
<tasktype>CodeletObjectDetector</tasktype>
<param name="object" type="string">good,no</param>
<param name="distance" type="int">1</param>
<param name="node"
type="string">negative_opinion</param>
</task>
```

Fig. 2. An example of configuration of codelets determining polarity of opinions.

Figure 2 presents an example of configuration of codelets determining polarity of opinions. Task name parameter (in LIDA codelet architecture is configured as a task, such as refreshing GUI content, which is performed by a codelet) defines the name of a codelet, task type indicates the name of a Java class in which codelet program code is contained, object parameter defines which words (or expressions) are searched for by a codelet in the sensory memory, no object parameter determines which words (or expressions) cannot exist in a text (for example, an opinionis positive once there can be found the word "recommended" which is not in close-proximity of the word "not"), distance parameter determines the maximum distancebetween words or expressions searched for in the working memory in case searchedwords (or expressions) have been found.

4. Results of analysis, in the semantic network form, are transferred to the workspace (a current situational model is created). Figure 3 shows an example of results of an analysis of the following opinion about a Product1: "Recommended, feature1 and feature2 is good". Nodes are represented using the symbol of a big circle, while links are represented using arrows. Dots represent levels of links activation (associations may be determined with a certain level of probability). The network shows a situation in which the general opinion about a product, as well as opinions on particular features of a product are positive.

Figure 4 shows an example of results of an analysis of the following opinion "Recommended, feature1 is good, but feature2 is not good".

5. In the next step, the situational model is passed to the global workspace and from the procedural memory the following patterns of action are automatically selected: "saving results of opinion analysis into a data base" (noSQL type data base – analysis results – semantic network – are saving in the XML format) and "loading next opinion into the sensory memory". It is also possible to select "statistical

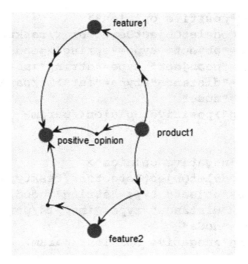

Fig. 3. An example of results of an analysis – positive opinion about the product and about the features.

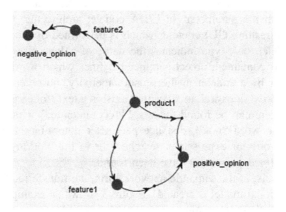

Fig. 4. An example of results of an analysis – positive opinion about the product and negative opinion about the one of the features.

analysis" on the basis of which an agent also highlights features which are most sought after by customers.

The following part of the paper contains results of verification of accuracy of customers' opinions analysis performed by a CRM cognitive agent.

The following part of the paper contains results of verification of accuracy of customers' opinions analysis performed by a CRM cognitive agent.

6   Research Experiment

In order to verify customers' opinions using the CRM cognitive agent, a research experiment has been carried out in which results of automatic analysis were compared with results of an analysis performed by a human (an expert), i.e. a manual analysis. The following assumptions were adopted in the experiment:

1. The analysis concerned opinions about TV sets from online shops, price comparison sites, and internet forums.
2. Shallow text analysis was performed (description of this category of analysis was shown in item 2 of the paper).
3. Number of analyzed opinions: 300. The limitation is connected with the fact that with respect to each single opinion, a manual analysis had to be performed, which is a time consuming procedure.
4. For the needs of the experiment, 13 features of TV sets were analyzed: brand, type, price, color, look, type of screen (LCD, LED, plasma), energy consumption, screen size, resolution, sound parameters, Wi-Fi technology, internet browser, types of slots (e.g. HDMI, USB, EURO, CI).
5. The assumption was that if an opinion did not contain information about polarity of a given feature the polarity of that feature was identical with the polarity of the opinion (for example by analyzing the following opinion "I recommend the TV set, its only drawback is poor sound" the polarity of the opinion was considered to be positive, the polarity of the "sound" feature was negative, and the polarity of remaining features was considered positive).
6. A method of learning with a teacher was employed. On the basis of a learning set consisting of 75 opinions a configuration of codelet parameters defining polarity of opinions and codelets of extraction of features and opinions about features was performed.
7. In order to determine the accuracy of results of automatic analysis in relation to results of manual analysis, the following measurements were performed:

   - effectiveness – this measure defines the relationship of the number of opinions whose polarity (or polarity of features) has been determined automatically to the number of opinions whose polarity (or polarity of features) has been determined manually; this measure enables one to determine in how many cases the polarity of opinions (or polarity of features) has not been determined by an agent (an agent has not specified whether an opinion (feature) is positive or negative); the next used measures relate only to opinions (features) effectively recognized by an agent,
   - precision – which specifies the accuracy of classification within a recognized class of opinions and it is defined in the following way

$$p = \frac{opp}{opp + onp} \tag{1}$$

where:

p – precision,

opp – positive opinions recognized as positive ones,

onp – negative opinions recognized as positive ones.

- sensitivity – the relationship of the number of opinions recognized by an agent as positive ones against all positive opinions is defined in the following way:

$$c = \frac{opp}{opp + opn} \tag{2}$$

where:

c- sensitivity,

opp – positive opinions recognized as positive ones,

opn – positive opinions recognized as negative ones.

- F1 measure – defined in the following way:

$$F1 = \frac{2 * (p * c)}{p + c} \tag{3}$$

where:

F1- F1 measure,

p – precision,

c – sensitivity.

All the presented measures have values ranging from 0 to 1.

Research experiment was carried out in the following way:

1. 300 randomly selected opinions found on online shop sites offering audio and video equipment, price comparison sites, and internet forums were recorded in a data base.

2. Manual analysis was performed (notes on the opinions).

3. Then, a learning set was created which contained 75 randomly selected opinions on the basis of which parameterization of codelets was made (learning with a teacher). The opinions were also grouped according to the degree of difficulty (four groups – group 1 – polarity of opinions and features easy to determine, group 4 - polarity of opinions and features difficult to determine) of determining their polarity and the polarity of features of TV sets characterized in the opinions (Table 1). The previously characterized measures of accuracy of results of analyses were calculated taking into consideration particular groups of degree of difficulty.

    Taking into consideration opinions of the first group it needs to be said that they contain only expressions concerning polarity of opinions (there is no information about individual features), it is easy to recognize the polarity, however polarity of individual features is treated as polarity of opinions. The second group comprises opinions containing information about polarity of individual features. It may be difficult to recognize a particular feature. The third group of opinions is characterized by ambiguous descriptions of polarity of individual features (in the sample opinion, the customer does not like the

Table 1. Types of opinions according the degree of difficulty of their analysis.

No.	Sample content of opinion (written in Polish language they was translated into English in this paper)
1	Recommended (not recommended), good (not good)
2	Recommended, advantages: - high quality of workmanship - price - value for money - meets expectations - look disadvantages: - poor sound quality I have had the TV set for about a month and I think it is really great, especially the picture which is clear and fluid (…) voice commands constitute a drawback.
3	The TV set is ok, the only thing that makes me angry is the frame, in photos it looks really nice and thin however once you turn the TV on it appears that the screen is smaller by additional circa 7 mm, which for me greatly spoils the effect. Generally, the TV is worth recommending.
4	I have bought the TV set for my mom. Previously she had a 3.5 years old Panasonic which did not work well with digital channels (problems with MPEG decoder) that's why I decided to replace it. However, me as well as my mom have noticed that picture quality of the Samsung TV set is far worse than Panasonic's, even though it has been calibrated. I wish I had paid a bit more and bought a Panasonic which has let me down a bit, but not as much as this particular Samsung.

Source: own work.

frame – negative opinion and at the same time he mentions that it is nice that the frame is so thin – positive opinion). The fourth group of opinions is the most difficult to analyze as opinions belonging to the group contain descriptions of several products in one opinion. It is difficult to determine which product should be assigned polarity of opinions and which product should be assigned features characterized in the opinion.

4. The next step involved loading, one by one, all the 300 opinions into the sensory memory, having them analyzed by an agent, and saving results of the analysis in a data base.

5. The last step involved calculation of measures of accuracy of automatic analysis results. Table 2 presents the list of obtained results.

Generalizing results of analyses of opinions about TV sets performed by a cognitive agent of the CRM sub-system one can state that efficiency, as well as precision and sensitivity of recognizing polarity of opinions is high which results from the fact that most often, while entering opinions, users have to give its polarity. Taking into account recognition of polarity of individual features one must say that efficiency of recognizing them is lower than the efficiency of recognizing polarity of opinions, which means that not all words (expressions) indicating polarity of a feature noted manually have been found by an agent. It resulted mainly from the fact that not all of the words (expressions) were present in opinions from a learning set. It needs to be noted that low values of precision measures, sensitivity and F1 with respect to polarity of features contained in opinions of group 4 (for example look, sound – value 0,5556 – marked in Table 2 with a dark-gray color) mean that the polarity of the features in many cases had not been recognized correctly, i.e. many features having positive opinion had been recognized as features of a negative opinion or just the opposite.

High efficiency of recognizing polarity of opinions of features such as "Brand", "Type", "Price", "Color" in the second or third group (value of measures is 1,0000 or is close to the value – marked in Table 2 with a light-gray color) results from the fact that in the opinions values of the features have been recorded in a structured form, however the opinions did not contain information about polarity of the features, which is why their polarity had been automatically determined according to the general polarity of opinion.

Table 2. Results of the opinions' and features' polarity analysis.

Group of the opinion	Measure	Polarity of opinions	Brand	Type	Price	Color	Look	Screen type	Energy consum	Screen size	Resolution	Source	Wi-Fi	Browser	Slots
1	Effectiveness	0,9697	0,9697	0,9697	0,9697	0,9697	0,9697	0,9697	0,9697	0,9697	0,9697	0,9697	0,9697	0,9697	0,9697
	Precision	0,9697	0,9697	0,9697	0,9697	0,9697	0,9697	0,9697	0,9697	0,9697	0,9697	0,9697	0,9697	0,9697	0,9697
	Sensitivity	1,0000	1,0000	1,0000	1,0000	1,0000	1,0000	1,0000	1,0000	1,0000	1,0000	1,0000	1,0000	1,0000	1,0000
	F1	0,9846	0,9846	0,9846	0,9846	0,9846	0,9846	0,9846	0,9846	0,9846	0,9846	0,9846	0,9846	0,9846	0,9846
2	Effectiveness	1,0000	1,0000	1,0000	1,0000	1,0000	0,8462	0,9615	1,0000	0,9615	1,0000	0,8077	0,8462	0,8846	0,8462
	Precision	1,0000	1,0000	1,0000	1,0000	1,0000	0,8750	1,0000	0,8750	0,9600	0,9600	0,8750	1,0000	0,8750	0,8261
	Sensitivity	1,0000	1,0000	1,0000	1,0000	1,0000	0,9545	0,9615	0,9545	0,9600	0,9600	0,9545	0,9615	0,9545	0,9048
	F1	1,0000	1,0000	1,0000	1,0000	1,0000	0,9130	0,9804	0,9130	0,9600	0,9600	0,9130	0,9804	0,9130	0,8636
3	Effectiveness	1,0000	1,0000	1,0000	1,0000	1,0000	0,8571	0,8571	0,9643	0,8571	0,8929	0,8929	0,8929	0,9286	0,7273
	Precision	1,0000	1,0000	1,0000	0,9600	0,9600	0,7727	0,9600	0,9600	0,9200	0,9200	0,6522	0,6522	0,7727	0,7727
	Sensitivity	1,0000	1,0000	1,0000	0,9600	0,9600	0,7391	0,9600	0,9600	0,8846	0,8846	0,6818	0,6818	0,7391	0,7391
	F1	1,0000	1,0000	1,0000	0,9600	0,9600	0,7556	0,9600	0,9600	0,9020	0,9020	0,6667	0,6667	0,7556	0,7556
4	Effectiveness	0,8462	0,7692	0,8462	0,8462	0,8462	0,7692	0,8462	0,8462	0,8462	0,8462	0,8462	0,8462	0,7692	0,6923
	Precision	0,9231	0,5556	0,5556	0,7778	0,7500	0,5556	0,7778	0,7500	0,7778	0,7778	0,5556	0,7500	0,7500	0,7500
	Sensitivity	1,0000	0,5556	0,5556	0,6364	0,9000	0,5556	0,6364	0,9000	0,6364	0,6364	0,5556	0,9000	0,9000	0,9000
	F1	0,9600	0,5556	0,5556	0,7000	0,8182	0,5556	0,7000	0,8182	0,7000	0,7000	0,5556	0,8182	0,8182	0,8182
Average	Effectiveness	0,9540	0,9347	0,9540	0,9540	0,9540	0,8606	0,9086	0,9450	0,9086	0,9272	0,8791	0,8887	0,8880	0,8089
	Precision	0,9732	0,8813	0,8813	0,9269	0,9199	0,7932	0,9269	0,8887	0,9069	0,9069	0,7631	0,8430	0,8419	0,8296
	Sensitivity	1,0000	0,8889	0,8889	0,8991	0,9650	0,8123	0,8895	0,9536	0,8702	0,8702	0,7980	0,8858	0,8984	0,8860
	F1	0,9862	0,8850	0,8850	0,9112	0,9407	0,8022	0,9063	0,9190	0,8866	0,8866	0,7800	0,8625	0,8678	0,8555

While analyzing average values of individual measures one can see that the highest recognizability has been obtained with respect to features "type", and "price", whereas the lowest effectiveness can be noted in case of the polarity of the feature "slot". While analyzing measures of precision, sensitivity, and F1 the highest value of the measures was obtained in case of the feature "price", whereas the lowest value was obtained in case of the feature "sound".

On the basis of opinions' analysis, the CRM agent stated also that customers prefer to buy TV sets with the following characteristics:

- price: 1200-2000 PLN,
- color: black,
- screen type: LED,

- low energy consumption,
- screen size 40–42 inch,
- resolution: Full HD,
- sound: Virtual Surround Sound,
- technology Wi-Fi: yes,
- internet browser: yes,
- slots: HDMI, USB, EURO, CI.

In the presented list, features such as brand, type or look have been omitted (opinions concerning looks have been characterized by a big diversity of values of the feature and it is impossible to state what is the most preferred look of a TV set among buyers).

Results of analysis performed by the CRM cognitive agent are available in real time, in the remaining sub-systems of a prototype of CIMIS. The results may for example be used in order to change production plans (production planning is performed by a cognitive agent of a production management sub-system) so as to increase the number of TV sets with features selected on the basis of analysis of customers' opinions.

The authors are aware of the fact that the corpus consisting of 300 opinions is insufficient to precisely determine effectiveness of the devised method of analyzing opinions, however it already indicates that the adopted method of solving the problem, after improvements, may prove effective. Works concerning carrying out an experiment on a corpus consisting of several thousands of opinions are under way.

7 Conclusions

The CIMIS as an application of computational collective intelligence allows for group decision supporting related with enterprise's processes management. Cognitive agents that operate in the system replace humans in making decisions on the operational, tactical and strategic level. They can also perform many routine activities instead of human (e.g. receiving an e-mails, actions related to production line). Of course, it is also necessary to apply appropriate actuators.

It is important to emphasize that such approach does not assume making employees redundant in the company, because, while the agent is performing a certain task, they should perform supervision over the agents, improve their knowledge and seek solutions that will expedite the operation of the company in a specific field. Therefore, the benefits from implementing the discussed system in a company will not be found in lower costs of employment, but rather in the following aspects: increasing work efficiency (the agent program can work non-stop while human work is connected with such events as unworked hours), having proper amount and most up-to-date information, drawing conclusions based on the information, accelerating the decision-making process (the agent makes decision in close-to-real time), lack of influence from non-substantive factors (such as fatigue, pressure from third parties) on the decisions made, increase in the automation of business processes, providing information and suggesting solutions to managers and employees, lower risk of work-related accidents (humans do not have to be present in the manufacturing hall).

Nowadays, integrated information management systems apart from processing structured knowledge must also enable processing unstructured knowledge (most often contained in text documents), which very often constitutes basis for taking decisions. Users' opinions that can be found in Internet resources are an example of the kind of knowledge. On the basis of analysis of the opinions one can take various decisions such, for example concerning planning production (or sales) of products with features which have been specified by customers in their opinions as positive ones. One of the methods of analyzing such opinions, in integrated information management systems, is using cognitive agent programs. As opposed to other methods, on the basis of results analysis, the programs are capable of taking decisions, which are automatically put into practice (for example deciding about starting production of new type of TV sets with features most desired by customers). Results of the research experiment carried out in the paper using the prototype of CIMIS system, have helped to draw a conclusion that the cognitive agent accurately recognizes polarity of opinions, and polarity of features of products in case of opinions in which the polarity has been unequivocally stated. In case of opinions characterized by ambiguous descriptions of polarity of individual features, or opinions containing specifications of several products in one opinion, the agent had difficulties in determining polarity of the products' features. Consequently, it is necessary to continue further research which will increase the accuracy with which an agent detects polarity of products' features.

Further work may require introducing changes in the functioning of the algorithm of codelets, as well as their configuration. Under way are also works aimed at implementing into the structure of LIDA cognitive agent methods of carrying out in-depth analysis of texts.

Acknowledgement. This research was financially supported by the National Science Center (decision No. DEC-2013/11/D/HS4/04096).

References

1. Better execute your business strategies – with our enterprise resource planning (ERP) solution, 28 Apr 2014. http://www.sap.com/pc/bp/erp/software/overview.html
2. Bytniewski, A. (ed.): Architektura zintegrowanego systemu informatycznego zarządzania. Wydawnictwo AE we Wrocławiu. Wrocław (2005) (in Polish)
3. Cognitive computing research group, 20 Apr 2014. http://ccrg.cs.memphis.edu/
4. Duch, W., Oentaryo, R.J., Pasquier, M.: Cognitive architectures: where do we go from here? In: Wang, P., Goertzel, P., Franklin, S. (eds.) Frontiers in Artificial Intelligence and Applications, vol. 171, pp. 122–136. IOS Press, Amsterdam (2008)
5. Davenport, T.: Putting the enterprise into the enterprise system. Harvard Business Review, pp. 121–131 (1998)
6. Franklin, S., Patterson, F.G.: The LIDA architecture: adding new modes of learning to an intelligent, autonomous, software agent. In: Proceedings of the International Conference on Integrated Design and Process Technology. Society for Design and Process Science, San Diego, CA (2006)

7. Goertzel, B.: OpenCogPrime: a cognitive synergy based architecture for embodied general intelligence. In: Proceedings of ICCI-2009 (2009)
8. Hecht-Nielsen, R.: Confabulation Theory: The Mechanism of Thought. Springer, Heidelberg (2007)
9. Hernes, M., Nguyen, N.T.: Deriving consensus for hierarchical incomplete ordered partitions and coverings. J. Univ. Comput. Sci. **13**(2), 317–328 (2007)
10. Hernes, M., Matouk, K.: Knowledge conflicts in business intelligence systems. In: Proceedings of Federated Conference Computer Science and Information Systems, pp. 1253–1258. Kraków (2013)
11. Hensinger, A., Thome, M., Wright, T.: Cougaar: a scalable, distributed multi-agent architecture. In: IEEE International Conference on Systems, Man and Cybernetics (2004)
12. Hofstadter, D.R., Mitchell, M.: The copycat project: a model of mental fluidity and analogy-making. In: Hofstadter, D., Fluid Analogies Research Group (eds.) Fluid Concepts and Creative Analogies, Chapter 5. Basic Books, New York (1995)
13. Joachims, T.: Text categorization with support vector machines: learning with many relevant features. In: Nédellec, C., Rouveirol, C. (eds.) Machine Learning: ECML-98. LNCS, vol. 1398, pp. 137–142. Springer, Heidelberg (1998)
14. Katarzyniak, R.: Grounding modalities and logic connectives in communicative cognitive agents. In: Nguyen, N.T. (ed.) Intelligent Technologies for Inconsistent Knowledge Processing, pp. 21–37. Advanced Knowledge International, Adelaide, Australia (2004)
15. Laird, J.E.: Extending the SOAR cognitive architecture. In: Wang, P., Goertzel, P., Franklin, S. (eds.) Frontiers in Artificial Intelligence and Applications, vol. 171, pp. 224–235. IOS Press, Amsterdam (2008)
16. Banaszak, Z., Klos, S., Mleczko, J.: Integrated management systems. Management and engeeniering of manufacturing. Polskie Wydawnictwo Ekonomiczne, Warszawa (2011)
17. Nazemi, E., Tarokh, M.J., Djavanshir, G.R.: ERP: a literature survey. Int. J. Adv. Manuf. Technol. **61**, 999–1018 (2012)
18. Nguyen, N.T.: Inconsistency of knowledge and collective intelligence. Cybern. Syst. **39**(6), 542–562 (2008)
19. Nguyen, N.T.: Metody wyboru consensusu i ich zastosowanie w rozwiązywaniu konfliktów w systemach rozproszonych. Wroclaw University of Technology Press (2002) (in Polish)
20. Pham, L.V., Pham, S.B.: Information extraction for Vietnamese real estate advertisements. In: Fourth International Conference on Knowledge and Systems Engineering (KSE), Danang (2012)
21. Plikynas, D.: Multiagent based global enterprise resource planning: conceptual view. Wseas Trans. Bus. Econ. **5**(6), 372–382 (2008)
22. Rohrer, B.: An implemented architecture for feature creation and general reinforcement learning. In: Workshop on Self-Programming in AGI Systems, Fourth International Conference on Artificial General Intelligence, Mountain View, CA, 11 Apr 2014. http://www.sandia.gov/rohrer/doc/Rohrer11ImplementedArchitectureFeature.pdf
23. Sebastiani, F.: Machine learning in automated text categorization. ACM Comput. Surv. (CSUR), New York **34**(1), 1–47 (2002)
24. Singhal, A.: Modern information retrieval: a brief overview. Bull. IEEE Comput. Soc. Tech. Comm. Data Eng. **24**(4), 35–43 (2001)
25. Sobieska-Karpińska, J., Hernes, M.: Consensus determining algorithm in multiagent decision support system with taking into consideration improving agent's knowledge. In: Proceedings of Federated Conference Computer Science and Information Systems, pp. 1035–1040 (2012)
26. Soderland, S.: Learning information extraction rules from semi-structured and free text. Mach. Learn. **34**(1–3), 233–272 (1999)

27. Sołdacki, P.: Zastosowania metod płytkiej analizy tekstu do przetwarzania dokumentów w języku polskim. Praca Doktorska Politechniki Warszawskiej, Warszawa (2006). (in Polish)
28. Tomassen, S.L.: Semi-automatic generation of ontologies for knowledge-intensive CBR. Norwegian University of Science and Technology (2002)
29. Tran, C.: Cognitive information processing. Vietnam J. Comput. Sci. 1(4), 207–218 (2014). Springer, http://link.springer.com/article/10.1007/s40595-014-0019-4
30. Trandabăţ, D.: Using semantic roles to improve summaries. In: Proceedings of the 13th European Workshop on Natural Language Generation (ENLG 2011), Association for Computational Linguistics, Stroudsburg, PA, USA, pp. 164–169 (2011)
31. Vlas, R.E., Robinson, W.N.: Two rule-based natural language strategies for requirements discovery and classification in open source software development projects. J. Manag. Inf. Syst. 28(4), 11–38 (2012)
32. Wang, P.: Rigid flexibility. The Logic of Intelligence, vol. 34. Springer, Netherlands (2006)
33. Wawer, A.: Mining opinion attributes from texts using multiple kernel learning. In: IEEE 11th International Conference on Data Mining Workshops (2011)
34. Wilson, T., Wiebe, J., Hoffmann, P.: Recognizing contextual polarity: an exploration of features for phrase-level sentiment analysis. Comput. Linguist. 35(3), 399–433 (2009)
35. Xpertis – intelligents systems of enterprise manufacturing, Macrologic, 15 Apr 2014. http://www.macrologic.pl/rozwiazania/erp
36. Zenkin, A.: Intelligent control and cognitive computer graphics. In: IEEE International Symposium on Intelligent Control, pp. 366-371, Montreal, California (1995)
37. Zhang, C., Zhang, X., Jiang, W., Shen, Q., Zhang, S.: Rule-based extraction of spatial relations in natural language text. In: International Conference on Computational Intelligence and Software Engineering (2009)

Agreements Technologies - Towards Sophisticated Software Agents in Multi-agent Environments

Mirjana Ivanović[(✉)] and Zoran Budimac

Department of Mathematics and Informatics, Faculty of Sciences,
University of Novi Sad, Novi Sad, Serbia
{mira,zjb}@dmi.uns.ac.rs

Abstract. Agreements are one of vital social concepts that help human agents to facilitate interactions in social settings. Nowadays, a challenging interdisciplinary scientific research includes all the processes and mechanisms concerned in reaching agreements between different kinds of agents. Enhancing agents with "social" abilities is newest trend in application of agent technology and the implementation of wide range of multi-agent systems.

Agreement Technologies bring a new flavor in implementation of more sophisticated autonomous software agents that mutually negotiate in order to come as much as close to win-win situation and to acceptable agreements.

The goal of the paper is to present key concepts in this area and highlight influence of Agreement Technologies on development of more sophisticated multi-agent systems. Several interesting systems and environments from different domains are briefly presented.

Keywords: Agreement technologies · Software agents · Multiagent systems · Semantics in agreements

1 Introduction

It is impossible to imagine contemporary world without agreements. Human ability to reach agreements is present in all their interactions and without them there is no cooperation in social settings. Human social skills represent an intriguing challenge for researchers and have led to the emergence of a new research field, Agreement Technologies (AT) [36, 37]. Agreement Technologies are connected to a wide range of computer systems in which autonomous software agents negotiate usually on behalf of humans to reach mutually acceptable agreements.

As a consequence, different interesting research questions in artificial intelligence arise, like: is it possible to build *computers* with social skills? Is it possible to develop software systems that will *reach agreements* on behalf of humans through cooperation and coordination?

In meanwhile a lot of high-quality research activities and initiatives emerged and significant scientific results are achieved in this area. Among the most important initiatives in the area of AT was a big COST Action project IC0801. The Action was funded for 4 years (2008–2012), comprised about 200 researchers from 25 European countries and 8 institutions from other continents. They worked on topics related to AT

© Springer-Verlag Berlin Heidelberg 2015
N.T. Nguyen (Ed.): Transactions on CCI XVIII, LNCS 9240, pp. 105–126, 2015.
DOI: 10.1007/978-3-662-48145-5_6

[29]. The overall mission of the project was to support and promote the harmonization of high-quality research towards a new paradigm for the next generation distributed systems. These systems have to be based on the notion of agreement between computational agents and to support technology transfer to industry. Project has been successfully finished and resulted in numerous theoretical and practical achievements and contributions. Challenges and opportunities for further research and high-quality contributions are numerous.

The rest of the chapter is organized as follows. In Sect. 2, basic concepts of Agreement Technologies are briefly presented. Section 3 brings wider view on basic concepts of AT and their role in multi-agent environments. Section 4 explains essential functionalities of several distributed systems that are highly based on AT. Last section concludes the chapter.

2 Agreement Technologies

In diverse working environments supported by a wide range of software applications, specific software components - *agents* support people in achieving their interests. Such systems caused shifting a paradigm from strict and centralized client-server architectures, towards more decentralized way of interaction.

Software agents are crucial software elements of open distributed systems characterized by:

- **autonomy**: agents operate without the direct intervention of humans or others, and have some kind of control over their actions and internal state;
- **social ability**: Using some agent-communication language agents interact with humans and other agents;
- **reactivity**: agents perceive their environment and respond timely to changes that occur in it;
- **proactiveness**: except of simple actions as response to their environment, agents can exhibit goal-directed behavior and take some kind of the initiative.

Interactions between agents and environment must be also supported by sophisticated activities like reasoning, learning, or planning.

In future open distributed systems interaction of *software agents* (or more complex *intelligent computational agents*) will be deeply based on concept of *agreement* and will include a normative model and an interaction [30]. In case of such essentially new kinds of systems additional elements are necessary and have to be considered:

- a normative context containing rules of the game, i.e. agreements that the agents can reach;
- an interaction mechanism where agreements must be first established, and after that enacted [30].

Interactions between sophisticated software agents can be abstracted to the establishment of *agreements for execution*, and a subsequent *execution of agreements*.

Agreements accordingly have to be changed dynamically at run-time and there must be mechanisms for re-assessing and revising them during the execution. A term "interaction-awareness" is introduced, where software components explicitly represent and reason about agreements and associated processes. There are several key dimensions where new mechanisms and solutions for the establishment of agreements need to be developed [3]: *Semantics*, *Norms*, *Organizations*, *Argumentation*, and *Trust* (Table 1).

Table 1. Key dimensions of agreement technologies [3]

Dimension	Challenges for further research innovations
Semantics	Application-dependent ontology support: domain-specific objects and language interpretation
	Explicit and exploitable representation of environment aspects and semantics
	Combining knowledge in large-scale open settings and reconciling subjective views
	Learning the semantics of everything, out of cases of inspecting and exploiting the specific interactions with the environment
	Creating and specifying commonly agreed languages for interaction
Norms	Standardization for representing the actions and events in an environment
	Representation standardization of the interactions context. Definition (in a standard way) of all necessary resources, properties, and appropriate values
	Invention and adjustment of adequate mechanisms for contextualizing abstract norms that appeared in design-time into norms that form specific spaces
	Easy way to add functionalities/services required for norms management usually extending those provided by the environment
Organization	Providing facilities to enter or exit a given organization. Essentially supporting run-time recruitment of new members and exclusion of existing ones
	Allowing basic activities of organizations on-demand: creation, deletion, and modification
	Obtaining adequate support to the institutional components of an organization (norms, powers, agreements)
	To fulfill all mentioned functionalities, agents usually use the artifacts of the environment. Develop adequate mechanisms to support it
Argumentation	It is necessary to have good mechanisms for scaling up typical, single interactions between a small number of agents
	Developing reliable mechanisms for managing libraries and database of ontologies, agreements, and protocols
	Agents in a system have to be able to reason about ontologies and activate adequate interaction protocols
Trust	Exploiting the powerful connection between trust/reputation and the environment in electronic settings
	It is necessary to have environment that is more trust and reputation friendly. Therefore, trust and reputation mechanisms have to influence the agent's actions in order to modify the environment in a friendlier manner

3   Key Dimensions of Agreement Technologies

As we already mentioned experts from AT area recognize and distinguish several important key dimensions that characterize AT: Semantic Technologies, Norms, Organizations and Institutions, Argumentation and Negotiation, and Trust. All of them will be explained and presented in more details in the following subsections.

3.1   Semantics in Agreement Technologies

In the recent period Web became rather matured and included variety of standards authorized by the World Wide Web consortium (W3C):

- **A uniform exchange syntax**: the eXtensible Markup Language (XML)
- **A uniform data exchange format**: the Resource Description Framework (RDF)
- **Ontologies**: RDF Schema and the Web Ontology Language (OWL)
- **Rules**: the Rule Interchange Format (RIF)
- **Query and transformation languages**: XQuery, SPARQL.

Within these standards, AT propose additional elements:

1. *Policies, Norms. And the Semantic Web "Trust Layer"* – Policies are in fact rules and constraints put in process of modeling behaviors. They include protocols to exchange policies and languages for rule specification with an intention to support describing and exchanging policies (as RIF - Rule Interchange Format and XACML - eXtensible Access Control Markup Language) (Fig. 1).

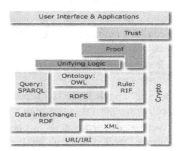

Fig. 1. Semantic web layers

Norms usually represent polices agreed in a community but also can be something individual. Policies on the (Semantic) Web also represent bases for privacy of organizational and personal data. Work on formalization of (community) norms and (private and organizational) policies are essential for a wide range of applications.

2. *Evolution of Norms and Organizational Changes* – Usually the evolution of organizational change and norms, and policies are connected to aligning and merging already existing norms and policies. Ontology languages based on Description Logics are not sufficient to express such semantic models and policies. So other formalisms

have to be considered for exploitation and "smooth" incorporation in multi-agent systems and environments.

3. *Organization-Based or Norm-Based Programming Languages* versus *Semantic Web Languages* – "trust layer" of the Semantic Web is still in very immature stage. Different languages and protocols (as P3P, XACML) have been constantly developing and assessing in prototypes and real systems. One among extremely important tasks in this area is oriented to find mechanisms for embedding rule based and formal descriptions of norms.

4. *Implicit Versus Explicit Norms on the Semantic Web* - Best practices and norms on the Web are not (yet) made explicit.

Logical formalisms for AT, i.e. Semantic Web standards serve for representing the knowledge of local agents. The main intention is to achieve a goal in agreement with other agents. In distributed, open, and heterogeneous systems that use AT, formalisms of Semantic Web suffer from various limitations. Autonomous agents define their knowledge according to their own beliefs. Semantic Web standards do not provide the means to compartment knowledge from distinct sources, so conclusions reached when using the global knowledge of disagreeing agents could be inconsistent.

Recently a number of logical formalisms invented and proposed to handle the situations appeared. They usually extend classical or Semantic Web logics [43] and the common name for them is *contextual logics* or *distributed logics* or *modular ontology languages* [22].

3.2   Norms in Agreement Technologies

Numerous agent environments and systems recently have been implemented using norms to obtain essentially higher quality. Norms become more important mechanisms to regulate electronic institutions and commerce. They also offer significant support for coordination and security in such systems.

Study of norms is interdisciplinary and it includes different forms and causes an innovative understanding of norms and their dynamics. Deontic logic, as a field of logic highly connected to norms, is concerned with obligation, permission, and related concepts. Also it is a formal mechanism for capturing the essential logical features of mentioned concepts. Deontic logic arises several essential and challenging research questions and dilemmas: Reasoning about Norm Violation, Normative Conflicts, Norm without Truth, Revision of a Set of Norms, Time and Action Issues, Knowledge and Intentions, Norm Emergence, and Games.

'BOID' architecture, proposed in [7], incorporates interaction between beliefs, desires, intentions, and obligations in the creation of agent goals. Essential issues discussed here include the interaction between 'internal' and 'external' motivations i.e. issues derived from the norms of the agent's social context.

Social and legal theories consider different types of norms like: **constitutive norms** that support 'institutional' actions like making of contracts, the issuing of fines; **regulative norms** describing obligations, prohibitions, and permissions. For normative reasoning in dynamic and uncertain environments constitutive norms play essential role

[5]. They basically support normative reasoning in dynamic and uncertain environments like realization of agent communication in electronic contracting.

Different simulations have been performed in early stage and works on application of norms and cooperation in software systems [4]. Study of social phenomena is emergent topic nowadays and interconnection between artificial intelligence and social sciences introduced new discipline oriented to multi-agent systems. Research in normative multi-agent systems is in expansion now and main assumption is that norms are specified by the institution and all the agents in the society know about these norms in advance [2]. Alternatively, researchers interested in the emergence of norms do not assume that agents know the norms in advance.

Recent research in this area is oriented on modeling agents interactions (coordination or cooperation) [35] and studying how norms emerge. Agents are supposed to perform few actions (e.g. cooperate and defect) and research is oriented on studying mechanisms that support these actions that an agent is capable of performing. In [34] authors propose an interesting approach that facilitate data-mining for the identification of norms. Quantity of domain knowledge and prior knowledge about norms an agent possesses play important role in norm identification.

All three well known aspects of active learning (learning based on doing, observing, and communicating) are important for software agents and current simulation-based works on norms. So in research community it is necessary to pay attention to deeply consider and facilitate these learning aspects. Most current studies that investigate norm emergence using simulations based on simple games only use learning based on doing. Having in mind limitations of this type of learning and necessity to employ other learning types it is unavoidable that in future research authors should integrate usage of these three types of learning in wide range of applicable domains.

In multi-agent systems that recently appeared, new research approach provides agents with the ability to identify the presence of norms through rewards and sanctions. A promising research direction for the study of norms could be an inclusion of 'humans in virtual agents' systems. It would be interesting to observe different simulations where agents can learn from human agents and software agents can recommend norms to humans.

3.3   Organizations and Institutions in Agreement Technologies

Open multi-agent systems in synergy with Agreement Technologies represent promising technologies for solving complex problems in different organizations and institutions. Using declarative specifications to a number of agents numerous complex tasks or problems in organizations can be solved. Agents organized in teams solve delegated task in order to reach the global goals in an organization. In addition, the notion of institution is used in agent communities to develop different socio-technical systems. Norm compliance could be ensured during the interaction among autonomous agents.

Organizational perspective proposes that the joint activity inside multi-agent systems regulated by formally specified plans, norms, structures, and/or mechanisms will achieve appropriate tasks. This is in fact somehow expected by an organizational perspective. An organizational model is based on a conceptual framework i.e. Organization

Modeling Language (OML). Organizational specification in such a model can be enacted on some organization management infrastructure or on a traditional multi-agent platform [15, 25]. Furthermore, using such organizational specification agents have to know how to access the specific infrastructural services and how to create specific requests. Such agents are *organization aware* and are able to observe the organization and decide if to enter or change such a structure, and whether or not to comply with the different rules existing in organization.

In contemporary complex socio-technical systems it is almost impossible to have updated all the information about the environment. In [38] authors recognize agent-oriented modeling as a holistic approach for developing organizations based on two kinds of agents: technical (artificial) and human. Human and artificial agents are active components performing a range of activities in environment, perceive events, and reason [38] in socio-technical organizations.

Several different organizational models have been proposed and implemented in last years. Moise (Model of organization for multi-agent systems) is an interesting example of organizational model presented in [24]. This model is based on Organization Management Infrastructure and Organization Modeling Language that facilitate development of organizational aware agents.

Also there exist some other interesting approaches and proposed organizational models: AGR [16], TAEMS [28], ISLANDER [15], OperA [13], AGRE [16], MOISEInst [17], ODML [23], TEAM [39], AUML [31], MAS-ML [12]. In Table 2 different modeling dimensions for these models are presented. This general analysis recognizes four cohesive categories of constructs for modeling in an organizational model:

- *Organizational Structure:* encompasses different time invariant constructs for representing aspects of the structure of the agent organization;
- *Organizational Functions:* encompasses constructs for specifying global goals and goal decompositions in agent organization;

Table 2. Organization modeling dimension for particular models

Model	Structure	Interaction	Function	Norms	Environment	Evolution	Evaluation	Ontology
AGR	+	+	−	−	−	−	−	−
TAEMS	−	−	+	−	+	−	+	−
ISLANDER	+	+	−	+	−	−	−	+
OperA	+	+	+	+	−	−	−	+
AGRE	+	+	−	−	+	−	−	−
MOISEInst	+	−	+	+	−	+	−	−
ODML	+	−	−	−	−	−	+	−
STEAM	+	−	+	−	−	−	−	−
AUML	+	+	+	−	+	−	−	−
MAS-ML	+	+	+	+	+	−	−	−
Moise	+	−	+	+	−	+	−	−
VOM	+	+	+	+	+	−	−	+
Agent-oriented	+	+	+	±	+	−	−	−
AAOL	+	±	±	+	+	+	±	−

- *Organizational Interactions:* encompasses constructs for specifying time-dependent aspects of standardized actions and interactions which are part of organizational structure and function;
- *Organizational Norms:* encompasses constructs to regulate how above mentioned organizational elements are interrelated.

Apart from mentioned widely present dimensions, there exist also other elements that are important for agent organizations:

- *Organizational Environment:* encompasses constructs to represent a collection of resources in the space of the agent organization;
- *Organizational Evolution:* encompasses constructs to model changes in the organization (norms, goals) and achieve adaptation to new demands from the environment in agent organization;
- *Organizational Evaluation:* encompasses constructs for measuring the performance of norms and formal structure in an agent organization;
- *Organizational Ontologies:* regarding the application domain of the agent organization, constructs from this category have to build conceptualizations that must be consistently shared by the component agents.

3.4   Augmentation and Negotiation in Agreement Technologies

Before inclusion of agreement in Agreement Technologies, this concept initially has been studied in philosophy and law. In the last decade it has been becoming extremely important in computing, especially for inferences/dialogs and systems for decision making/support/negotiation. Generally, argumentation focuses on interactions where different stakeholders plead for and against some conclusions. They play important role when incomplete and inconsistent information appeared in systems. They are also helpful in situations where resolution of conflicts and diversity of opinion between different parties appeared. Agreement as a valuable mechanism to reach some consensus also benefits from negotiation. It is especially important when autonomous agents have conflicting interests/desires but may benefit from cooperation.

The theory of argumentation is interdisciplinary research area that includes: philosophy, communication studies, linguistics, psychology, and artificial intelligence.

Argumentation is naturally predominantly modular and majority of formal theories of argumentation adopt the following three things: (1) arguments are constructed in some underlying logic; (2) interactions between arguments are defined; (3) given the network of interacting arguments, the winning arguments are evaluated.

Potential for implementation of logical models of argumentation has been illustrated in recent research and also wide range of their application in different software systems appeared. Additionally any complex process that ensures an agreement assumes some kind of conflict and necessary steps and techniques for its resolution. Such conflicts may arise among different parties involved in variety of negotiating situations and dialogs (like the reasons or arguments for offers, stated beliefs, or proposed actions). For example in e-commerce systems (online negotiations involves automated software agents) in a handshaking protocol, a seller would simply successively make offers that have been either accepted or rejected. The exchange of arguments provides agreements that would not be reached in simple handshaking protocols.

Having these facts in mind it is clear that argumentation may be of significant value in sophisticated software systems based on agreement technologies.

3.5   Trust and Reputation in Agreement Technologies

Methods of computational trust and reputation have reached significant level of maturity at the moment. The multi-agent systems as a rather new paradigm initiated an evolution of some topics explored in this area. Trust and reputation mechanisms are not black-box and isolated processes that agents perform. It is a good time to join different parts and processes that were studied independently in developing and implementation of complex software systems. Computational trust and reputation processes have to be joined with other different parts and processes of the agents to facilitate execution and realization of complex tasks and mutual activities. They especially have to be connected to the notion of interaction and existence in an environment.

In humans interactions the trust is present in a day-to-day routine. In every person-to-person interaction a certain kind of decision about trust has to be made. Such interaction connected or not to an action is known as a "choice with commitment" [11]. Also, it always denotes an agent (trustee) behavior that may interfere with the truster own goals.

Trust has a vital role in society and it is a challenging research topic in multidisciplinary milieu that include disciplines like sociology, philosophy, social psychology, economics, political science, and management, but recently also computer science and especially multi-agent systems [29]. Intelligent agents improve their social interactions by capability to estimate the trustworthiness of interacting partners [33]. In fact, using trust theories, such agents are equipped by *computational trust models* in order to improve their trust-based decisions. Variety of notions and concepts of trust theory helps in revealing a "degree of confusion and ambiguity that plagues current definitions of trust" [9]. So computer scientists are able to easily formalize models of computational trust in artificial entities and enable them to make higher quality decisions.

Trust is not necessarily mutual or reciprocal and could be considered twofold, first as a decision and not an act, and second as a multi-layer concept that includes disposition and decision [9]. *Degree of trust* is also very important trust concept. Measure of the degree of trust is the *strength of trust* (i.e., the confidence that the truster has on his trust).

In [14] authors introduced concept of situational trust specifying trust as "a measurable belief that the truster has on the competence of the trustee in behaving in a dependable way, in a given period of time, within a given context, and relative to a specific task".

Understanding how trust forms and evolves is necessary to construct robust computational trust models. This will allow intelligent agents to promote their own trustworthiness, and to correctly predict others' trustworthiness even in case of new partnerships.

Reputation is also a social concept as complex as trust. Interrelation between these two concepts is rather ambiguous: reputation is an antecedent of trust, and it may or may not influence the trust; or on the other hand the process of reputation building is subject to specific social influences. Therefore trust and reputation could be seen as isolated constructs - thus reputation *does not* influence trust.

Recently several computational trust models have been proposed in the distributed artificial intelligence with intention to allow agents to make trust-based decisions. Majority of them primarily used the aggregation of past evidence about the agent under evaluation with the aim to estimate its trustworthiness. In spite the fact that recent models process specific properties of trust and its dynamics, still there is a long path to run in computational trust.

Although computational *reputation* raised a set of specific research questions, different researchers have proposed their own models of computational trust and reputation. These models integrate both social concepts, assuming the perspective of reputation as an antecedent of trust [26, 33].

4   Real World Applications

In the last decade a lot of research groups and authors have been working on implementation of different frameworks, architectures, and environments that demonstrate the use of Agreement Technologies in a variety of real-world scenarios [1, 18, 20, 21, 42]. This section briefly discusses some of these systems.

4.1   A Multi-agent Decision Support Tool for Water-Right Markets

Paper [18] presents multi-agent decision support tool called mWater, as an open and regulated virtual organization in which intelligent agents try to manage a flexible water-right market [6].

The proposed system has to be used as a simulator and facilitator in decision-making processes for policy makers. It focuses on two issues:

- on demands and specifically on the type of regulatory that includes norms selection and agents behavior,
- on market mechanisms to prevent possible conflicts between involved parties in achieving an efficient use of water.

Water insufficiency is a major interest and care of authorities in most countries. It is a consequence of: uncertain balance in types of use, conflicts over evaluation of water needs, and water rights. Some general considerations are that better organization, functioning, and effects could be obtained within an institutional and decentralized framework. Within such framework, usually fulfilling some in-advance set of norms, water rights could be exchanged freely to other users. Initial intentions of the authors have been to incorporate in the system different social aspects (first of all intelligent agents equipped with trust, cooperation, argumentation i.e. Agreement Technologies) in order to achieve a win-win situation in efficient use of water resources.

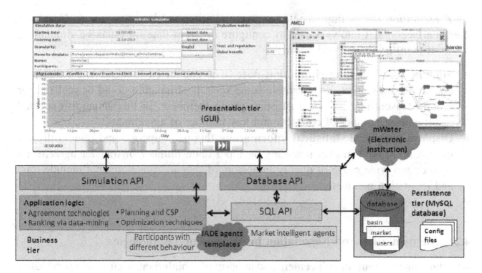

Fig. 2. Multi-tier architecture of the mWater system [18]

A flexible water-right market called mWater has been designed with two primary intentions. First to *deploy a virtual market* and observe interplay of agents, rule enforcing, and performance indicators in decision-support tool. Second intention was to provide a *playground* for the agreement computing paradigm which supports easily plugged-in new techniques and assess their impact in the market indicators.

Proposed system is based on a multi-tier architecture and an electronic institution model (Fig. 2) that includes five agents' roles.

- **guests** - potential users, before really entering the market,
- **water users** - guests that have valid water rights,
- **buyer/seller** - water user that currently joins for the market,
- **third parties** - water users, directly/indirectly affected by a water transfer (usually conflicting parties).
- **market facilitator and basin authority** - governing roles of the market.

Agreement execution may be conflicting with third party agents so system has to solve normative conflicts within the institution using grievance structure.

Architecture of the system is organized in multi-tier manner that include Persistence, Business, and Presentation tiers.

Persistence tier is in fact relational database that includes entire information about basins, markets, and grievances.

Business tier is an essential component that uses different and supportive artificial intelligence techniques like: Trust and data mining for participants' selection, planning to navigate through the institution, and Collaboration/negotiation to enhance agreements and minimize conflicts. Deliberative module is part of staff agents and his main duties are to support: reasoning on regulation matters, simulation how regulations and norms modify the market behavior and finally evaluation and analyses of their effects.

Presentation tier is implemented in very intuitive and interactive way. Graphical user interface - GUI allows a user to configure different data for effective simulation. Some of them are mentioned below:

- **Starting and finishing simulation date**: it is connected to water users that participate in the market. For example some specific simulation could be realized for group of users that do not trust other group members. In this situation expected result is high number of conflicts and low number of agreements.
- **Displays graphical statistical information**: this part supports and offers different information about market reactions on input data; number of transfer agreements signed; and wide range of quality indicators of "social" functions that include trust and reputation parameters.

4.2   Augmentation and Negotiation in Agreement Technologies

An interesting system for call centers and support of a high quality customer services in such centers is presented in [21]. In fact such system employs case-based argumentation and allows the technicians working in a call centre to provide a high quality customer support by reaching adequate agreements. Main motivation of authors to develop such system is based on the fact that nowadays, a lot of companies offer very similar products, prices and quality. Their business objective is to try to be better than their competitors by offering focused customer care supported by high quality and fast service within call centre.

On the other hand on the labor market, the less experienced technicians are cheaper and for employers it is more interesting to hire them than more expensive and experienced technicians. It seems that a better solution for employers is to provide less experienced technicians and to solve (collaboratively) as many requests as possible with arguing, contrasting views with other technicians, and reaching agreements [21]. Customer support offered by the company could be improved using case-based reasoning on previously stored and later reused the final solution applied to each problem.

In the previous version of the system, agents were allowed to use their case-bases to give experience-based support. Better effects could be achieved by integration of all technicians' knowledge in a unique case-based reasoning module. But this solution would probably need data mining of extra large case-bases and could be costly and complex. To solve this complexity an appropriate solution could be to have a unique but distributed case-based reasoning mechanism. But again it is not too realistic as it require that all technicians are willing to share their knowledge unselfishly.

In the new version of the system [21] the technicians are implemented as software agents. Agents are highly engaged in an argumentation process. They try to find the best solution to be applied to each new incidence received by call centre. This hybrid system integrates an argumentation framework where agents are equipped with argumentation capabilities and individual knowledge resources. In such multi-agent systems, software agents are capable to manage and exchange arguments taking into account the agents' *social context* (roles, dependency relations, and preferences).

Virtual call center is in fact society of agents that acts on behalf of a group of technicians that must solve problems in a Technology Management Centre (TMC). Agents' roles in virtual call centre corresponds to real technicians roles, so authors distinguish operator, expert, and administrator agents. Also, each agent can have its own values that it wants to promote or demote (adjust the reasons that an agent has to give preference to certain decisions) and that represent its motivation to act in a specific way.

The system functions in the following way. When a new request is received appropriate *ticket* is generated representing description of the problem to be solved. Often in the system can appear complex case where a ticket must be solved by cooperation of several agents (as technicians). They must argue to reach an agreement over the best solution to apply. Each agent possesses its own knowledge resources. They help him to generate a solution for the ticket. To solve the problem they apply appropriate data-flow for the argumentation process (see Fig. 3):

1. The system is organized as a group of technicians, i.e. agents that have to solve a new situation/ticket.
2. Using an argumentation module and applying the case-based argumentation each technician generates his own solution.
3. All technicians participating in the argumentation process are aware in each moment of the proposed positions.
4. The technicians willing to participate argue to reach the most suitable agreement.
5. The best solution is proposed to the user and feedback is registered in appropriate way in the system with intention to be used in future similar situations.

The system has been implemented and tested in a real call centre. The Magentix2 agent platform (http://users.dsic.upv.es/grupos/ia/sma/tools/magentix2/index.php) has been used for implementation of this multi-agent system. Magentix2 provides new services and tools that allow the secure and optimized management of open multi-agent systems.

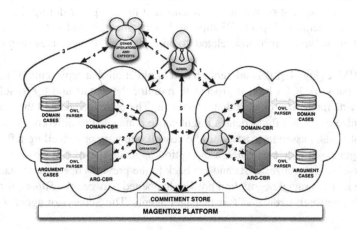

Fig. 3. Data-flow for the argumentation process of the call centre application

It has been integrated as an argumentation module that agents can use to persuade other agents to accept their proposed solutions as the best way to solve the problem.

4.3   ANTE: Agreement Negotiation in Normative and Trust-Enabled Environments

Another interesting usage of agreements in B2B electronic contracting is presented in [8]. The authors presented the ANTE framework that is based on three main agreement technology concepts: negotiation, normative environments, and computational trust. In spite of the fact that ANTE is research project oriented to B2B electronic contracting, it also could be applied as a more general framework in wider range of domains. It highly supports multi-agent collective coexistence resulting in two important activities and processes. First, it is negotiation as facilitator for achieving mutually acceptable agreements. Second, it is the enactment of agreements and the evaluation of the enactment phase, in order to improve future negotiations.

Each negotiation participant has its own contribution to the overall solution (determined by agreement). "It is therefore useful to represent the outcome of a successful negotiation process in a way that allows for checking if the contributions of each participant do in fact contribute to a successful execution of the agreement" [8]. "Computational trust may therefore be used to appropriately capture the trustworthiness of negotiation participants, both in terms of the quality of their proposals when building the solution" [8].

Computational trust may therefore be used to appropriately capture the trustworthiness of negotiation participants, both in terms of the quality of their proposals when building the solution. Computational trust adopted in ANTE framework is based on the ability to compute adequate estimations of trustworthiness in several different environments, including those of high dynamicity.

ANTE has been applied in different scenarios and an interesting application in disruption management in Airline Operations Control Centre (shortly AOCC) is briefly presented here. The AOCC is the organization responsible for monitoring working environment and solving operational problems that might appear during the execution of the airline operational plan. Disruption Management includes teams of experts specialized in solving problems related to aircrafts and flights, crewmembers and passengers.

MASDIMA (Fig. 4 [8]) is an agent-based application that represents AOCC of an airline company. AOCCs have a process to monitor the events and try to solve optimally the wide range of problems that may occur. The main idea behind these activities is to minimize costs that flight delays might cause connected to different aspects: crew costs, flight costs, passenger costs, and the cost of delaying or cancelling a flight from the passenger point of view. When a disruption appears, the AOCC needs to find the best solution that minimizes costs and get back to the previous operational plan and try to improve it. The crucial negotiation process happened between the *Supervisor* and the *A/C* and also between *Crew and Pax* manager agents. The Supervisor agent acts as the

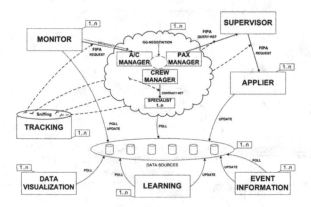

Fig. 4. MASDIMA architecture

organizer agent and on the other hand the manager agent acts as respondent. Each individual manager does not possess the full expertise to be able to propose a proper solution to the supervisor. So manager needs to activate an inter-manager negotiation to be able to complete their proposal and participate in the main negotiation. To minimize activities it is enough to have at least one manager for each part of the problem to take care about. Nevertheless, in the environment more than one agent with the same expertise in the same dimension of the problem can exist.

In this scenario, trust is used when the supervisor is evaluating negotiation proposals from the managers. The trust information is built in process of application of the winner solution on the environment through the *Applier* agent. Applier agent is responsible to apply the winning solution to the environment and also to check if execution of the solution has been successful. Supervisor agent enables connection of monitoring facility with the trust engine to use various evidences regarding the quality of the solutions.

To present an adequate proposal the manager agents first need to find appropriate candidate solution using a wide range of resources that exist on the operational plan. To find these candidate solutions, each manager might have a team of problem solving agents that are able to find and propose those solutions.

4.4   The Augmentation Web

The next example that illustrates practical use of Agreement Technologies is Argument Web presented in [27]. Argument Web is the plethora of argument visualization and mapping tools testified to the enabling function of argumentation-based models for human clarification and understanding; and for promoting rational reasoning and debate. The development of such tools was emerged as a consequence of existence of enormous number of discussion forums on the web, and on the other hand the lack of support for checking the relevance and rationality of online discussion and debate. One of extremely important characteristics of such tools is that they offer possibility to reuse

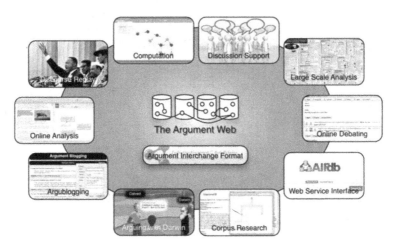

Fig. 5. The Argument web

readymade arguments authored online (i.e. with extensive mining of arguments from online resources).

To fully facilitate the development and employment of such tools and the reuse of authored arguments, new systems and standards on the Internet are necessary and needed. This leads to the completely new concept Argument Web [32] (Fig. 5). It can be seen and served as a common platform that brings together applications in different domains (e.g. broadcasting, mediation, education and healthcare) and interaction styles (e.g. online argument analysis, real-time online debate, blogging).

Online infrastructure for argument is combined with specific software tools that make interaction with the argument web easy and intuitive for various audiences. To support this appropriate standard for argument representation is needed - Argument Interchange Format or AIF [10].

A number of examples of specific interactions with the Argument Web that illustrate usage of prototype tools have been developed at the School of Computing, University of Dundee. More details and characteristics about them could be found in [29].

In the rest of the section a particular example of usage of Argument Web so called *Argumentation and the Social Web* will be briefly illustrated.

Primary functionality and aim of Social Web is to support users to connect with each other and share their knowledge and experiences of all types. Often they debate with frequent exchange of arguments (e.g. in comments on blogs). However, the argumentative structure is implicit, arguments need to be inferred, and debates are unstructured, often even chaotic.

Whereas the use of argumentation in the Social Web context has been advocated by some authors [40], the realization of such a vision is still far away from reality.

Most recent work on online systems and argumentation focuses on extracting argumentation frameworks using argument schemes and semantic web technology for editing and querying arguments [32]. It implicitly assumes that the extraction of argumentation frameworks is down to "argumentation engineers", fluent in computational argumentation.

So, for example if you want in modern search engines to obtain answer on some questions, you have to just type a couple of specific keywords. Usually as an answer you get a number of links that somehow satisfy your question, and you could be lucky if several first links bring information you needed. In contemporary information society the way people access the Web is rapidly and essentially changing. Among other activities a lot of people from highly developed societies are functioning every day intensively using and changing their status in Facebook or Twitter. A possible reason of this change in the Internet life style is that Facebook friends will actually give each other better information and through such communication decrease useless spam. Even more than that, search engines are interested in higher-quality opinion mining and sentiment analysis [19].

It is expected that future use of the web will be based on sentiment-aware search engine that mines large online discussion boards, include advanced clustering of results based on user agreement/sentiment, and fully integrates argumentation in a Social Web context.

4.5   Agreement Technologies in Learning Environments

Nowadays business learning practice and informal education are important areas for employing and using different learning technologies. It is interesting to notice and comprehend educational benefits related to work-place learning. Also another important direction in both formal and informal education is to identify associated problems to be solved with the combination of technologies and learning practices. Almost every company nowadays is facing an increasingly changing world and needs adaptation of learning mechanisms on all levels: individual, organizational, and group. The whole organization needs to be working together and learning together. It is a collective as well as individual learning activity and experience.

Our motivation [41] in recently started research topics and activities was to consider possibilities and opportunities to use Agreement Technologies for different learning purposes. At the beginning we try to propose a general conceptual learning framework which will encompass and exploit some aspects of Agreement Technologies in informal learning and help in solving previously unseen situations and complex problems.

For successful companies it is necessary to constantly adapt their roadmaps, innovations, and strategies to changes in their environments.

Action roadmapping assumes active involvement of the company's employees/ external stakeholders in strategic planning, implementation of innovations identified in roadmaps. Also roadmapping becomes an unavoidable learning process for the organization. On the other hand, dynamic roadmapping form the business practice point of view and implies an iteration of the company's roadmaps - as new information, opportunities, and threats. Also it assumes a set of technologies and tools to enable collaborative strategic planning, and foresight based future analysis. Agreement Technologies could be considered as one among other learning technologies to help in different situations where individuals or groups are facing unexpected situations/problems. In such situations they need more additional information, hints, and suggestions from variety of other sources. These sources can be colleagues from the same company, other people from local and global networks, available knowledge bases that have stored descriptions of similar cases and applied appropriate solutions and so on. Information, suggestions, and advices from mentioned sources could help in solving new situation based on someone previous experiences in the same or similar situations. As majority of sources could be previously unknown for employees they somehow have to trust them and be sure that given advices and suggestions are not malicious. One of possible coarse-grain scenarios for getting appropriate trustworthy hints and directions for understanding and solving such situations could be:

- Find experts/co-innovators (internally/externally) that can help the roadmapping groups build the innovations.
- Social intelligent agents/tools have to be trustworthy and conform to norms, with ability of case-based reasoning:

  - enable adequate description of the problems,
  - be able to search across the internet/inside the company's knowledge bases or/and those of its collaborators who possess adequate knowledge and skills to solve this particular unseen situation,
  - search for people who have the expertise to help or they had faced with similar problems before and could offer possible solution,
  - from the questions to an agent using appropriate (formal) language, agent uses case-based reasoning to bring back solutions/suggestions,
  - above mentioned process is repeated/iterated (including different parties) until the best satisfactory solutions are found.

The main question in this environment is "Can employee trust the quality of the agent's solution?" To obtain as much as possible trustworthy solutions and hints for new situation, it is necessary to employ Agreement Technologies and other intelligent agents' capabilities.

Simplified version and early conceptual learning model that considers Agreement Technologies is presented in Fig. 6.

Important feature of the system is that agents can trust each other. For this purpose similar idea presented in [20] could be used. Authors proposed LocPat, a personalized recommendation approach that provides recommendations to a requester in an agent network based on trust. Authors showed that the similarity score can be seen as an indicator of how much the agent can be trusted by the requester. Similarity score (in fact has been translated into trust value predictions) has been calculated as the similarity between the agent network and a structure graph. Important characteristic of this system that could be extremely important in our environment is that LocPat provides flexible personalized recommendation from the perspective of an agent.

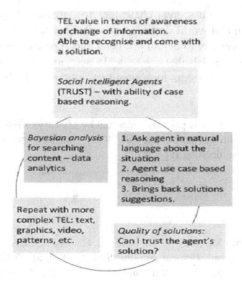

Fig. 6. Simplified learning scenario for unexpected situations based on AT

5 Conclusions

The paper brings some key concepts, dilemmas, and possible usage of Agreement Technologies in open distributed environments predominantly based on multi-agent systems. These define environments that are based on norms, argumentations, and trust within which agents interact. Agreement Technologies are obviously contemporary, interesting, and promising research area. Its multidisciplinary and interdisciplinary character offers great future possibilities for applications in more intelligent and sophisticated artificial societies. Practically there are no limitations where Agreement Technologies could be applied. There is wide range of domains like: E-Commerce, Transportation Management, E-Governance, Smart energy grids, highly technologically supported learning environments and so on.

In the E-Commerce domain possible applications could be in B2B transactions to support interactions in a dynamic and adaptive manner. In B2C transactions scenarios for customer support and product guidance could be supported by AT.

For example in domain of Transportation Management, self-interested behaviour of agents (drivers, passengers, etc.) could be organised by a set of norms (traffic rules).

Agreement Technologies could find their proper place and usage within different methods and tools for modeling, simulating, and evaluating processes and policies (citizens, public administrations) in E-Governance area.

Finally smart energy grids and virtual power plants could achieve new, higher and essential quality and better performance by employing AT. In the future energy grids, thousands of small-scale producers of renewable energy (solar, wind, etc.) could be distributed across the transmission/distribution networks and – taking into account certain norms and regulations – may decide to act together temporarily as *virtual power plants* [32].

Furthermore AT introduces great challenges for large-scale open distributed systems. But to use AT in real world scenarios, practical bounds to these techniques need to be removed. It is necessary to obtain test beds to compare different approaches, and also to propose efficient models that can scale. In networks of thousands or millions of interoperating services and agents high quality solutions to scale up agreements are needed.

Acknowledgements. The work is partially supported by Ministry of Education and Science of the Republic of Serbia, through project no. OI174023: "Intelligent techniques and their integration into wide-spectrum decision support"

References

1. Alberola, J.M., Such, J.M., Botti, V., Espinosa, A., Garcıa-Fornes, A.: A scalable multiagent platform for large systems. Comput. Sci. Inf. Syst. J. **10**(1), 51–77 (2013)
2. Aldewereld, H., Dignum, F., García-Camino, A., Noriega, P., Rodríguez-Aguilar, J.A., Sierra, C.: Operationalisation of norms for usage in electronic institutions. In: Proceedings of the Fifth International Joint Conference on Autonomous Agents and Multi-agent Systems (AAMAS), pp. 223–225. New York, ACM (2006)
3. Argente, E., Boissier, O., Carrascosa, C., Fornara, N., McBurney, P., Noriega, P., Ricci, A., Sabater-Mir, J., Schumacher, M.I., Tampitsikas, C., Taveter, K., Vizzari, G., Vouros, G.A.: Environ. Agreem. Technol. AT **2012**, 260–261 (2012)
4. Axelrod, R.M.: The evolution of cooperation. Basic Books, New York (1984)
5. Boella, G., van der Torre, L.W.: Constitutive norms in the design of normative multiagent systems. In: Toni, F., Torroni, P. (eds.) CLIMA 2005. LNCS (LNAI), vol. 3900, pp. 303–319. Springer, Heidelberg (2006)
6. Botti, V., Garrido, A., Giret, A., Noriega, P.: An electronic institution for simulating waterright markets. In: Proceedings of the III Workshop on Agreement Technologies WAT@IBERAMIA (2010)
7. Broersen, J., Dastani, M., van der Torre, L.: Beliefs, obligations, intentions and desires as components in an agent architecture. Intern. J. Intell. Syst. **20**(9), 893–920 (2005)
8. Cardoso, H.L., Urbano, J., Brandão, P., Rocha, A.P., Oliveira, E.: ANTE: agreement negotiation in normative and trust-enabled environments. In: Demazeau, Y., Müller, J.P., Rodríguez, J.M., Pérez, J.B. (eds.) Advances on PAAMS. AISC, vol. 155, pp. 261–264. Springer, Heidelberg (2012)
9. Castelfranchi, C., Falcone, R.: Trust Theory: A Socio-Cognitive and Computational Model. Wiley Series in Agent Technology. Wiley, Chichester (2010)
10. Chesñevar, C., McGinnis, J., Modgil, S., Rahwan, I., Reed, C., Simari, G., South, M., Vreeswijk, G., Willmott, S.: Towards an argument interchange format. Knowl. Eng. Rev. **21**, 293–316 (2006)
11. Cohen, P.R., Levesque, H.J.: Intention is choice with commitment. Artif. Intell. **42**(2–3), 213–261 (1990). doi:10.1016/0004-3702(90)90055-5
12. da Silva, V.T., Choren, R., de Lucena, C.J.P.: A UML based approach for modeling and implementing multi-agent systems. In: Proceedings of the Autonomous Agents and Multi-agent Systems, International Joint Conference on 2, Los Alamitos, IEEE Computer Society, pp. 914–921 (2004)

13. Dignum, V.: A model for organizational interaction: based on agents, founded in logic. Ph. D. Thesis, Universiteit Utrecht (2004)

14. Dimitrakos, T.: System models, e-risks and e-trust. In: Proceedings of the IFIP Conference on Towards the E-Society: E-Commerce, E-Business, E-Government, I3E2001, pp. 45–58. Deventer, Kluwer (2001)

15. Esteva, M., Rodríguez-Aguilar, J.A., Rosell, B., Arcos, J.L.: AMELI: An agent-based middleware for electronic institutions. In: Proceedings of the Third International Joint Conference on Autonomous Agents and Multi-agent Systems (AAMAS), pp. 236–243. New York, ACM (2004)

16. Ferber, J., Gutknecht, O., Michel, F.: From agents to organizations: an organizational view of multi-agent systems. In: Giorgini, P., Müller, J.P., Odell, J.J. (eds.) AOSE 2003. LNCS, vol. 2935, pp. 214–230. Springer, Heidelberg (2004)

17. Gâteau, B., Boissier, O., Khadraoui, D., Dubois, E.: Moiseinst: An organizational model for specifying rights and duties of autonomous agents. In: Third European Workshop on Multi-agent Systems (EUMAS 2005), Brussels, pp. 484–485 (2005)

18. Giret, A., Garrido, A., Gimeno, J.A., Botti, V., Noriega, P.: A MAS Decision Support Tool for Water-Right Markets. In: Yolum, P., Tumer, K., Stone, P., Sonenberg, L., (eds.) Proceeding of 10th International Conference on Autonomous Agents and Multi-agent Systems (AAMAS 2011), 2–6 May 2011, Taipei, Taiwan, pp. 1305–1306 (2011)

19. Godbole, N., Srinivasaiah, M., Skiena, S.: Large-scale sentiment analysis for news and blogs. In: Proceedings of the International Conference on Weblogs and Social Media (ICWSM), Salt Lake City (2007)

20. Hang, C.W., Singh, M.P.: Generalized framework for personalized recommendations in agent networks. Auton. Agent. Multi-Agent Syst. 25(3), 475–498 (2012)

21. Heras S.: Case-based argumentation framework for agent societies. Ph.D. Thesis, Departamento de Sistemas Informáticos y Computación. Universitat Politècnica de València (2011) http://hdl.handle.net/10251/12497

22. Homola, M.: Distributed description logics revisited. In: Proceedings of the 20th International Workshop on Description Logics DL'2007, 8–10, June 2007, Brixen/Bressanone, Italy (2007) http://ceur-ws.org/Vol-250/paper_51.pdf

23. Horling, B., Lesser, V.: A survey of multi-agent organizational paradigms. Knowl. Eng. Rev. 19(04), 281–316 (2005)

24. Hübner, J.F., Sichman, J., Boissier, O.: A model for the structural, functional, and deontic specification of organizations in multi-agent systems. In: Bittencourt, G., Ramalho, G.L. (eds.) Advances in artificial intelligence. LNCS, vol. 2507, pp. 439–448. Springer, Heidelberg (2002)

25. Hübner, J.F., Boissier, O., Kitio, R., Ricci, A.: Instrumenting multi-agent organisations with organisational artifacts and agents. Journal of Autonomous Agents and Multi-Agent Systems 20(3), 369–400 (2010)

26. Huynh, T.D., Jennings, N.R., Shadbolt, N.R.: An integrated trust and reputation model for open multi-agent systems. Auton. Agent. Multi-Agent Syst. 13, 119–154 (2006)

27. Kirschner, P.A., Buckingham Shum, S.J., Carr, C.S.: Visualizing Argumentation: Software Tools for Collaborative and Educational Sense-Making. Springer, London (2003) http://oro. open.ac.uk/12107/

28. Lesser, V., Decker, K., Wagner, T., Carver, N., Garvey, A., Horling, B., Neiman, D., Podorozhny, R., NagendraPrasad, M., Raja, A., Vincent, R., Xuan, P., Zhang, X.: Evolution of the gpgp/taems domain-independent coordination framework. Autonomous Agents and Multi-Agent Systems, vol. 9, no. 1, pp. 87–143, Kluwer Academic Publishers Hingham, USA (2004)

29. Ossowski, S.: Agreement Technologies, Springer Series: Law, Governance and Technology Series, vol. 8, no. XXXV, p. 645 (2013)
30. Ossowski, S., Sierra, C., Botti, V.: Agreement Technologies: A Computing perspective, Springer Series: Law, Governance and Technology Series, vol. 8, pp. 3–16 (2013)
31. Van Dyke Parunak, H., Odell, J.J.: Representing Social Structures in UML. In: Wooldridge, M.J., Weiß, G., Ciancarini, P. (eds.) AOSE 2001. LNCS, vol. 2222, pp. 1–16. Springer, Heidelberg (2002)
32. Rahwan, I., Zablith, F., Reed, C.: Laying the foundations for a worldwide argument web. Artif. Intell. **171**, 897–921 (2007)
33. Sabater-Mir, J., Paolucci, M.: On representation and aggregation of social evaluations in computational trust and reputation models. Int. J. Approx. Reason. **46**(3), 458–483 (2007)
34. Savarimuthu, B.T.R., Cranefield, S., Purvis, M.A., Purvis, M.K.: Obligation norm identification in agent societies. J. Artif. Soc. Soc. Simul. **13**(4), 3 (2010). http://jasss.soc. surrey.ac.uk/13/4/3.html
35. Sen, S., Airiau, S.: Emergence of norms through social learning. In: Proceedings of the Twentieth International Joint Conference on Artificial Intelligence (IJCAI), pp. 1507–1512. AAAI Press, Menlo Park (2007)
36. Shaheen Fatima, S., Wooldridge, M., Jennings, N.R.: Optimal Negotiation of Multiple Issues in Incomplete Information Settings. AAMAS 2004, pp. 1080–1087
37. Sierra, C., Botti, V., Ossowski, S.: Agreement computing. Künstliche Intelligenz **25**(1), 57–61 (2011)
38. Sterling, L., Taveter, K.: The art of agent-oriented modeling. MIT, Cambridge/London (2009)
39. Tambe, M., Adibi, J., Alonaizon, Y., Erdem, A., Kaminka, G.A., Marsella, S., Muslea, I.: Building agent teams using an explicit teamwork model and learning. Artif. Intell. **110**(2), 215–239 (1999)
40. Torroni, P., Gavanelli, M., Chesani, F.: Arguing on the semantic grid. In: Argumentation in Artificial Intelligence, pp. 423–441. Springer (2009) doi: 10.1007/978–0-387-98197-0_21
41. Vesin, B., Klašnja, M.A., Ivanović, M., Klašnja, M.A., Budimac, Z.: Ontology-based architecture with recommendation strategy in java tutoring system. Comput. Sci. Inf. Syst. J. **10**(1), 237–261 (2013)
42. Vrdoljak, L., Podobnik, V., Jezic, G.: Forecasting the acceptance of new information services by using the semantic-aware prediction model. Comput. Sci. Inf. Syst. J. **10**(3), 1025–1052 (2013)
43. Zimmermann, A.: Integrated distributed description logics. In: Proceedings of the 20th international workshop on description logics DL'2007, Brixen/Bressanone, Italy, 8–10 June 2007, pp. 507–514. Bolzano University Press (2007). http://ftp.informatik.rwthaachen.de/ Publications/CEURWS/Vol-250/paper_37.pdf

Identification of Underestimated
and Overestimated Web Pages Using
PageRank and Web Usage Mining Methods

Jozef Kapusta[✉], Michal Munk, and Martin Drlík

Constantine the Philosopher University in Nitra, Tr. A. Hlinku 1,
Nitra 949 74, Slovakia
{jkapusta,mmunk,mdrlik}@ukf.sk

Abstract. The paper describes an alternative method of website analysis and optimization that combines methods of web usage and web structure mining - discovering of web users' behaviour patterns as well as discovering knowledge from the website structure. Its primary objective is identifying of web pages, in which the value of their importance, estimated by the website developers, does not correspond to the real behaviour of the website visitors. It was proved before that the expected visit rate correlate with the observed visit rate of the web pages. Consequently, the expected probabilities of visiting of web pages by a visitor were calculated using the PageRank method and observed probabilities were obtained from the web server log files using the web usage mining method. The observed and expected probabilities were compared using the residual analysis. While the sequence rules analysis can only uncover the potential problem of web pages with higher visit rate, the proposed method of residual analysis can also consider other web pages with a smaller visit rate. The obtained results can be successfully used for a website optimization and restructuring, improving website navigation, and adaptive website realisation.

Keywords: Web usage mining · Web structure mining · PageRank · Support · Observed · Visit rate · Expected visit rate

1 Introduction

The aim of the website designers and developers is to provide information to users in a clear and understandable form. Web pages are mostly understood as an information resource for users. The hyperlinks interconnect information on web pages.

The website developers affect the visitors' behaviour by creating links between web pages. They indicate the importance of information displayed on web pages through the references. More references route probably to the more relevant web pages. They are often directly accessible from the homepage or are referred from other important web pages.

These web pages can also provide valuable information in the opposite direction. The website developers can obtain easily interesting information about the users of the website or users' behaviour, needs or interests.

© Springer-Verlag Berlin Heidelberg 2015
N.T. Nguyen (Ed.): Transactions on CCI XVIII, LNCS 9240, pp. 127–146, 2015.
DOI: 10.1007/978-3-662-48145-5_7

The aim of this paper is to explore the differences between the estimated importance of web pages (obtained by the methods of web structure mining) and visitors' actual perception of the importance of individual web pages (achieved by the methods of web usage mining). Moreover, we can use these differences between the expected probability and observed probability of accesses to the different web pages for identifying suspicious web pages at the same time. We defined the suspicious web pages as web pages that are not ordered correctly in the hypertext structure of the website for the purpose of this paper. In other words, the suspicious web pages are underestimated or overestimated by the website developer.

The approach used in this paper combines the findings and methods of two contemporary research areas, web usage mining (WUM) and web structure mining (WSM). The WUM methods assume that a user usually posts enough amount of information to the server during his visit to the web page. The web server usually saves this information in the form of log file records, which can be analysed in detail.

On the other hand, WSM methods focus on the analysis of the quality and importance of the web pages based on the references, which are directed to them [1]. The determination of the web page importance is based on the idea that the degree to which we can rely on the web page quality is transferred by references to the web pages. If the web page is referred to by other relevant pages, the references on that page also become relevant.

This paper is organised as follows: the second section provides a review of the related research in the field of WUM and WSM. The third section deals with the used methodology. The next section is focused on the results of a case study in which a residual analysis was applied in the context of WUM and WSM. The last section discusses the obtained results and provides several open questions for future research.

2 Related Work

Much research in recent years has focused on analysis of users' behaviour using web mining techniques and methods. However, only several researchers tried to combine web structure mining, web content mining and web usage mining methods. The combination of methods and techniques of these research fields could help to solve some typical issues of the web structure analysis and improve the personalisation, website structure and adaptivity tasks.

For example, Liu et al. [2] utilised learning algorithms for web page quality estimation based on the content factors of examined web pages (the web page length, the count of referred hyperlinks).

Chua and Chan [3] dealt with the analysis of selected properties of examined web pages. They combined the content, structure and character of hyperlinks to the web pages for their classification. They intended to find more precise web page classifiers for thematically oriented search engines. Each analysed web page was represented by a set of properties related to its content, structure and hyperlinks. Finally, they introduced 14 features of web pages, which they used consequently as the input of the machine learning algorithms.

Jacob et al. [4] designed an algorithm WITCH (Webspam Identification Through Content and Hyperlinks), which combined a web structure and web content mining methods for the purpose of spam detection. We also found a similar approach in other experiments [5–8].

Lorentzen [9] found only a few studies, which combine two sub-fields of web mining. The research was focused on a combination of the usage and content mining methods, but he also mentioned some examples of structure mining, which could be said to be web mining's equivalent to the link analysis. For instance, the Markov chain-based Site Rank and Popularity Rank combined structure and usage mining with a co-citation-based algorithm for the automatic generation of hierarchical sitemaps of websites, or for the automatic exploration of the topical structure of a given academic subject, based on the HITS algorithm, semantic clustering, co-link analysis and social network analysis.

Usually, the estimation of the web page quality was assured by the PR, HITS or TrustRank algorithms. However, low quality, unreliable data or spam in the hypertext structure caused less efficient estimation of the web page quality [10, 11]. Jain et al. [12] provided a detailed review of PR algorithms in web mining, their limitations and a new method for indexing web pages.

Ahmadi-Abkenari [13] introduced a web page importance metric of LogRank that worked based on an analysis of the datasets on the different levels of server click-streams. The application of this metric assumed that the importance of each page was based on the observation period of log data and non-dependency from the downloaded portion of the Web.

Agichtein et al. and Meiss et al. [14, 15] used the traffic data to validate the PageRank Random Surfer Model. Su et al. [16] proposed and experimentally evaluated a novel approach for personalized page ranking and recommendation by integrating association mining and PageRank.

A website optimization and restructuring, improving website navigation, and adaptive website realisation represent the most frequent application domain of the WUM and WSM methods combination [17].

Shutong et al. noted that a relevant page clustering is one of the most important methods of website optimization. They mentioned a page clustering improvement algorithm combining website topological structure and web page contents to improve the interest of mining results [18].

Wen-long and Ye-zheng [19] proposed a novel website optimisation model based on users' access pattern. They stated that a successful website must be adaptive with the abilities of automatically improving its hyperlink structure by learning from users' access pattern. They analysed the usage of a website and its structure and recommended the modifications to the website structure as a reaction to the changes in access patterns of its users.

Jeffrey et al. presented the approach that is closely related to the approach described in this paper. They considered a number of visits and total time spent by all users on a web page for web page importance ranking. They suggested a novel approach to identifying problematic structures in websites. Their method compared user behaviour, derived via web log mining techniques, to an analysis of the website's link structure obtained by applying the Weighted PageRank algorithm [20].

Finally, Wang and Liu [21] focused their research on development of a framework for the realisation of adaptive web site. Their framework was based on the combination of web log mining and topology analysis.

3 Research Methodology

We mentioned previously that we tried to combine WUM and WSM methods for the purpose of identification of the differences between the expected and observed accesses to the web pages of a website, i.e. for identification of suspicious (overestimated and underestimated) web pages. We describe the details of the proposed methodology on the case study in this section.

3.1 Web Usage Mining

The analysis of users' behaviour represents the main objective of WUM [1, 22]. The web page visitors' segmentation [23] and discovering visitors' behavioural patterns belong to the most cited WUM application domains [24–27]. WUM is concerned with secondary data that is created during the web pages visits in contrast to WSM, which deals with primary data usage.

Secondary data is stored in the log files of the web servers or proxy servers, error log files, in the cookies of the internet browsers, or it is stored in the user profiles. While the information systems store data in their own internal structure in the databases, data about the website accesses is stored in standardized text form of the log files, referred to as Common Log File (CLF). The CLF contains information such as IP address, access date and time, referrer and reference object. In addition, the extended CFL automatically adds information about the web browser version stored in a variable User-Agent.

Data Pre-processing in WUM. High quality of stored data is a fundamental prerequisite for successful data analysis. Poor quality of data at the input stage leads to imprecise analysis results regardless of the applied analytical method. This assumption is more serious in the case of WUM, where the analytical methods require careful pre-processing. Pre-processing is the most time-consuming phase of WUM because available data can be imprecise and incomplete, or it can contain non-essential information.

For example, the request for displaying a particular web page also contains a request to send the necessary CSS styles, JavaScript files and all images (.png, .gif, .jpg). Information about these files is subsequently recorded in the log file of the web server. The final log file, for this reason, contains not only GET requests, but also other requests of HTTP protocol.

We must clear the log file and remove such unnecessary data at the pre-processing phase. Paradoxically, this phase is considered the easiest phase of data pre-processing because it requires only data filtering using well-defined regular expressions.

Tracking web crawlers' visits to the web page represents the second source of unnecessary records stored in the log files, which we should remove. The crawlers browse the website sequentially, they behave differently to humans. For this reason, the records are not suitable for WUM analysis.

User Session Identification. The web server log file is a primary source of anonymous data about a user (a website's visitor). Anonymous data can cause at the same time a problem in uniquely identifying a web page visitor. Identifying particular users is not as vital as distinguishing between different users in WUM.

The problem is that the user can visit the web page repeatedly. For this reason, the web log file can contain multiple sessions of the same user. The objective of this phase of the pre-processing is the creation of discrete user relations. We call this process user session identification [28]. We can define these relations in several manners. According to Spiliopoulou [29] we define a user session as a sequence of necessary steps required to fulfil a particular task successfully. The second definition describes the session as a sequence of steps leading to a particular aim [30].

The user session identification method based on the different IP address of the website visitors represents the easiest approach. This approach has one disadvantage. The IP addresses are not suitable for individual user identification in general because several users can share the same IP address, for example, when they are connected behind the NAT (Network Address Translation) or proxy devices.

We can solve this problem by introducing the authentication of users, but its use must take into account data protection legislation [31]. We must also mention the next group of issues with user identification using IP addresses. When the user visits the website from several computers simultaneously, or the user uses different IP addresses or different web browsers on the same computer or other devices, we are not able to distinguish them. If the user takes an advantage of the available tools for data protection, the success rate of user session identification using IP addresses is low because these tools assign the user different IP addresses during one session.

We should note the alternative methods of user session identification briefly. These methods are based on the information stored in the cookies files saved in the text file on the user's computer. The first time the user visited the website, the server sent the cookies to the web browser along with the requested web page. If the user visited the website again, the web browser sent cookies to the server. For this reason, the web server was able to identify individual users.

It is necessary to emphasise the fact that the cookies are closely bound to the web browser. Each web browser manages its cookies. This means that cookies help us to distinguish distinct users that use the same computer in a classroom or Internet cafe. Of course, this assumption is correct only if each user uses his web browser and has own account. This method may not be accurate in some cases:

- Users can remove cookies stored on their computers and other devices; the server, therefore, will not be able to identify returning visitors [32].
- Some web browsers do not support cookies.
- Web browsers allow users to ban cookies [33].

Even though, cookies are considered the most common and the most simple method of user session identification, the issues above limit their practical use and have to be replaced by other methods [34].

For this reason, the session identification method using time-window represents the most common method. Using this method, each time we found subsequent records about the web page requests where the time of the web page displaying has been higher than explicitly selected time, we divided the user visits into several sessions. Explicitly chosen time was denoted as Standard Time Threshold (STT) and it may take different values: 5 min [35], 10 min [36], 15 min [37] or 30 min [38]. This method is widely used because of its simplicity.

Finally, we note other methods for completeness' sake. The first method introduced the dynamic STT [39, 40], another method used information about the web browser version [28] or heuristic methods with referrer or the site map [41]. The last method dealt with the value of H-ref [42].

Reconstruction of Users' Activities with Path Completion. The reconstruction of the website visitors' activities represents another issue of WUM. This technique does not belong to the session identification approaches. It is usually the next step of the data pre-processing phase [43]. The main aim of this phase is identifying any significant accesses to the webpage that are not recorded in the log file [44, 45].

For example, if the user returned to the web page during the same session, the second attempt to access probably displayed the cached version of the webpage. Tauscher and Greenberg [46] proved that the website users realised the access to a particular web page by clicking on the web browser Back button, context menu or by using mouse features in more than 50 % of cases. These actions were not recorded in the web server log files because there was not any request to the web server.

We assume that the web browser cache causes the aforementioned problems. The path completion technique provides an acceptable solution to these problems. The main idea of this technique assumes that we can reconstruct the missing records in the web server log file using the site map or eventually using the value of the variable referrer stored in the log file [47, 48]. If one or more previous web pages contain the hypertext links to the examined web page, we will complete the path to the web page from the nearest possible web page.

The Fig. 1 depicts the typical structure of a web page. The arrows indicate the user's transition between web pages. Dashed arrows represent the transition between web pages using the Back button or mouse gestures. We can see that the user returned from the web page D to the web page A through pages C and B. Subsequently, the user clicked on the hyperlink leading to page X. We can see the references to pages C, B and A are missing in the log file (Table 1), because they were already realised on the client side. It is clear that there is a discrepancy between the real activity of the user and the records stored in the log file.

We should be aware of one another important fact. If hyperlinks also exist from the web page B or C to the web page X, and we have only one log file, then there probably exist an unlimited count of possible path completions [49].

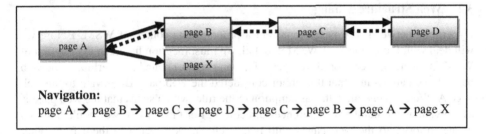

Navigation:
page A → page B → page C → page D → page C → page B → page A → page X

Fig. 1. User transition through the sequence of web pages.

Table 1. Log file records.

URL address	Referrer
page A	–
page B	page A
page C	page B
page D	page C
page X	page A

If we deal with the path completion in detail, we can not omit the importance of the sitemap. The sitemap represents the structure of the given website, and it contains information about the relations between web pages.

We can illustrate the usefulness of the sitemap on the sequence of visited web pages $A \rightarrow B \rightarrow C \rightarrow D \rightarrow X$ for a selected IP address. We created an algorithm that was able to find a missing hyperlink from the web page D to the web page X. We could assume the user clicked on the Back button from some of the previous web pages. We used a backtracking algorithm to find the first web page with the direct hyperlink to the web page X.

Considering this assumption we could assume in the given example that if web page C did not contain the reference to web page X, the proposed algorithm added the web page C to the previous sequence of visited web pages $A \rightarrow B \rightarrow C \rightarrow D \rightarrow C \rightarrow X$.

Similarly, we found the absence of the hyperlink between the web page B and web page X. Therefore, we added it to the sequence $A \rightarrow B \rightarrow C \rightarrow D \rightarrow C \rightarrow B \rightarrow X$.

In the next step, the algorithm found that web page A contained the hyperlink to the web page X.

Finally, we obtained the final sequence of visited web pages $A \rightarrow B \rightarrow C \rightarrow D \rightarrow C \rightarrow B \rightarrow A \rightarrow X$. This means that the user clicked on the Back button between web pages D and C, C and B, and also between web pages B and A.

It is evident that the proposed algorithm of a path completion had to be able to estimate the access time to the supplemented web page correctly at the same time. In this case, the calculation of average access time between directly unlinked web pages represents the easiest method that can be used.

3.2   Web Structure Mining

Hyperlinks between web pages represent a primary data source of WSM. Before we will deal with the methods of WSM in detail, we must mention the main assumption of WSM. If the author (creator, developer) of a web page A created a hyperlink to the web page B, we can assume that the author considered the web page B relevant to the web page A. We can generalize this assumption to the rule, which states that the web page is as relevant as the number of references direct to it. For this reason, we should assign a greater priority to the web pages with many references directed to them from other relevant web pages [49].

Information retrieval was the first application, which benefited from the web structure mining. The extra work involved in ordering of results was its main disadvantage.

In this section, we summarise the information about the most common algorithm PageRank and explain its role in the discovering of knowledge from the web structure mining.

The PageRank (PR) interprets the hypertext link from the web page X to the web page Y as a vote of the web page X for web page Y. PR represents not only the frequency of votes, but it also contains much more valuable information. The PR algorithm also analyses the web page that has given the vote. In general, the votes of the important web page have a higher weight than less important web pages [49].

We consider PR as a static evaluation of web pages i.e. PR is calculated for each web page off-line, independently of searching queries.

We divide the web page hyperlinks into two categories:

- In-links – all hyperlinks that refer to the web page i from other web pages.
- Out-links – all hyperlinks that refer to other web pages from the web page i.

The recursive hyperlinks are not considered. At the same time, we assume, the hyperlink from the web page, which referred to other web pages, transferred its importance to the target web pages implicitly. It means the web page is more relevant if other important web pages refer to it. We consider the web as an oriented chart

$$G = (V, E), \tag{1}$$

where V is a set of nodes, i.e. a set of all web pages and E is a set of oriented edges, i.e. hyperlinks among web pages.

Let n ($n = |V|$) be the total count of web pages of the website. Then the PR of the web page i ($Pr(i)$) is defined as

$$Pr(i) = \sum_{(j,i)\in E} \frac{Pr(j)}{O_j}, \tag{2}$$

where O_j is the count of hyperlinks referred (Out-links) to the other web pages from the web page j.

Brin and Page [50], the authors of the PR, introduced the Random Surfer Model. This model assumed that the user clicked on the hyperlinks, he/she never returned, and they have started on another random web page.

Therefore, the authors introduced Dumping Factor d, d $\langle 0,1 \rangle$. Dumping factor represented the probability that the random surfer would continue on the next web page. The value $1 - d$ meant the probability that the user would start on the new web page. The typical value of the variable d is usually 0.85 [50].

If we consider the set E defined in (2), the transformed equation for PR will be

$$Pr(i) = (1 - d) + d \sum_{(j,i) \in E} \frac{Pr(j)}{O_j}, \qquad (3)$$

where $0 \leq d < 1$ and O_j is the count of hyperlinks referred (Out-links) to the other web pages from the web page j.

We can iterate this calculation until the value of $Pr(i)$ begins to converge to the limit value [51]. The PR algorithm, which is currently used by the web browsers, has undergone several changes. We intended to find the importance of the individual web pages of the website in the experiment. For that reason, we decided to implement the original version of the PR algorithm, not its improved or more actual version.

4   Results

We could see input data of web mining from two different views:

- Data, which depends on the web page developers, this data represents the input of the WSM methods. We used them for PR calculation.
- Data, which depends on the web page visitors, this data represents the input of the WUM methods. We used them for calculation of visit rates of individual web pages.

These two views of data are interdependent. The web page developers should create web pages that reflect the needs of their visitors. The visitors' behaviour was determined by the structure and the content of the web page and vice versa, but we had the primary origin of data in mind.

4.1   Data Pre-processing for PR Calculation

We developed the crawler, which went through and analysed web pages of a selected university website. The crawler began on the homepage and read all hyperlinks on the examined web page. If the crawler found hyperlinks to the unattended web pages, it added them to the queue.

The crawler created a sitemap that we utilized later in the PR calculation of individual web pages. The web crawler implemented the method, which operated in several steps:

1. URL selection from the queue,

2. an analysis of the content of selected web page for the purpose of finding new URL references,
3. new URL references added to the queue.

The crawler was simple because it scanned only the hyperlinks between the web pages. We consider this as the main limitation of the proposed method, because the crawler did not regard the actual position of the hyperlinks within the web page layout, which has a strong influence on the probability of being accessed by a website visitor.

When the crawler finished, we created a hypertext matrix from the sitemap. Consequently, we calculated PR for individual web pages according to the formula (1). The dumping factor d was 0.85.

4.2 Data Pre-processing for Calculation of Visit Rate of Individual Web Pages

We used the log files of the university website. We removed unnecessary records and accesses of crawlers from the log file. The final log file had 573020 records over a period of three weeks. We prepared data in several steps:

1. Data with session identification using standard-time threshold (STT) [52, 53]. We used STT = 10 min.
2. Data with path completion [44, 45]. We used the method mentioned in Sect. 3. The number of records increased by 23 % and the mean length of the sequences was 6.
3. The observed visit rates of the web pages were found by the WUM method and represented by the value of variable *support*. The variable *support* is defined as $support(X) = P(X)$. In other words, item X has support s if s % of transactions contain X, i.e. the variable *support* means the frequency of occurrence of given set of items in the database. It represents the probability of visiting a particular web page in identified sequences (sessions).

4.3 Discovering Differences Between Expected and Observed Probability of Web Page Visits

We analysed the dependence of expected values of the web page visits on real values obtained from the log files with the aim to compare the importance of the web page given by the developer to the actual importance of this web page from the visitor's point of view.

The value of PR for individual web pages represented the probability of being visited by a random visitor. At the same time, PR expressed the importance of the web page from the web developer's perspective. If the web page was important, the developer created more references directed to this web page than to other, less important, web pages. We proved the dependence of values PR and real visit rate expressed as values of the variable *support* in [54].

Based on these findings, we also proposed an alternative method for discovering the differences between the expected and observed probability of accesses to the individual web page of the website. We found inspiration in the residual analysis.

The residual analysis serves for the purpose of the model validity verifying and its improving because it helps to identify the relationships, which the model did not consider. For example, we can use the residual analysis for a verification of regression model stability [55], i.e.; we can identify the incorrectness of the selected model using the correlated chart of residues and independent variable.

The main idea of residual analysis assumes that

Data = prediction using model (function) + residual value.

We used a pre-processed log file from a web server with users' sessions identified using the STT method. Simultaneously, we used calculated values of PR for each web page on the examined website. Finally, we used calculated values of the variable *support*. We only used the web pages with a value of variable *support* greater than 0.5 in the residual analysis.

We considered the values for the comparison of expected (PR) and observed (*support*) visit rate. The variable *support* was from the interval 0–100 and represented the probability of visiting a particular web page in the identified sessions.

We have been aware that the values of PR of individual web pages created together the probability distribution of all visits to the website. For this reason, the sum of PR should be 1. We transformed it into relative values for that reason.

The values of variable *support* represented the observed values in the described experiment. The values of the variable PR represented the expected values. As we mentioned earlier, the primary objective of the residual analysis was the identification of the outliers. We could visualise the residues in the charts of defined cases (Fig. 2).

After subtracting the values obtained from the model (expected values) from the observed values, we got errors (residual values). We could analyse the residual values for the purpose of the model appraisal.

The selection residues e_i are defined as

$$e_i = y_i - \hat{y}_i, \tag{4}$$

where $\hat{y}_i$ are expected values predicted by the model and y_i are observed values.

We created a chart with expected, observed and residual values for better understanding (Fig. 3). The chart visualises calculated values of residues, minimum, maximum, mean and median for each case.

The individual web pages ordered by the PR represent the x-axis. We can see the identifier of a web page in the chart of residues. We could also see from this graph that the expected values of PR and observed values of the variable *support* were different. The residuum had to be equal to zero in the ideal case, i.e., the expected and observed values should be the same. It implies; the structure of the references to a given web page created by the web developers would be better if the value of the residuum is closer to the x-axis.

Case	Residual		Residual
	-5,81	6,49	
1 (/university-structure/students-dormitories)			-5.4575926
2 (/about-the-university)			-4.6926729
3 (/study)			0.53558373
4 (/university-structure)			-1.1946187
5 (/about-the-university/contacts-list)			3.10384574
6 (/admissions)			5.9986803
7 (/continuing-education)			-4.9771971
8 (/research)			-5.8135952
9 (/media-ukf)			-5.5705865
10 (/university-events)			-5.6972031
11 (/information)			-2.9316893
12 (/component/user/login)			-4.8916465
13 (/admissions/electronic-return-receipt)			-0.705071
14 (/admissions/registration-to-study/191-registration-instructions)			-4.1334218
15 (/university-structure/faculty-of-education)			0.10730521
16 (/university-structure/faculty-of-arts)			0.75997713
17 (/university-structure/faculty-of-social-sciences-and-health)			-0.4209571
18 (/university-structure/fakulta-of-natural-sciences)			-0.469309
19 (/admissions/courses)			0.95291878
20 (/admissions/conditions-of-admission)			1.90343602
21 (/admissions/study-registration)			0.39091441
22 (/admissions/admission-results)			6.48515264
23 (/university-events/2996-nitra-summer-university)			0.43148715
24 (/direct-links/431-admissions)			0.57512837
25 (/informations/tenders)			1.04037486
26 (/direct-links/2479-academic-calendar)			2.62056342
27 (/continuing-education/current-offer)			0.24299822
28 (/study/study-organization)			0.22800527
29 (/study/135)			0.23563467
30 (/study/studentship)			0.24478994
31 (/study/134)			0.77732168
32 (/study/accredited-study-programmes)			1.6150292
33 (/admissions/electronic-return-receipt/2477-browser-setting-manual-)			0.3612422
34 (/documents/helpdesk/AIS/electronic-return-receipt-manual.pdf)			1.33933058
35 (/news/3000-6-6-2013-students-dormitories-committee-result-from-4-6-201-)			0.299168
36 (/university-structure/students-dormitories/1481-Zobor)			0.39072073
37 (/university-structure/students-dormitories/1480--Nitra)			0.43802298
38 (/admissions/conditions-of-admission/178)			2.25513976
39 (/admissions/study-organization/370)			0.56769477
40 (/admissions/study-organization/369)			0.68976508
41 (/admissions/study-registration/179-study-registration)			0.47301024
42 (/university-events/2995-meeting-faculty-of-educations-alumni)			0.64784142
43 (/news/2998-30-5-2014---DIPLOMA-2014)			1.34840738
44 (/direct-links/925-academic-year-schedule)			0.55961904
45 (/documents/university-structure/students-dormitories/04062013.pdf)			0.62826102
Min.			-5,8135952
Max.			6,48515264
Mean			-0,1935154
Median			0,43148715

Fig. 2. Chart of residues.

We did not consider the homepage of the website in this case study because the value of the variable *support* for this page was equal to 53.51. The average value of the variable *support* was 3.09. For this reason, the value of the home page would have distorted the residues of other web pages in the charts.

The main menu of the examined website caused the main problem of the described experiment. The main menu was available on each web page of the website. It means that there were always the direct references to the main parts of the website from all web pages. For this reason, there occurred the evident difference between the value of PR of web pages available directly from the main menu and other web pages (Fig. 3).

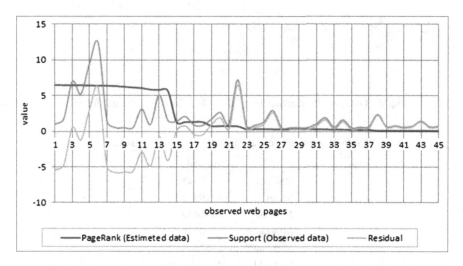

Fig. 3. Chart of expected, observed and residual values (Color figure online).

4.4 Residual Outliers Identification

The identification of outliers is the objective of a residual analysis. In this case study, the outliers identified potentially "suspicious" web pages. Precisely, the outliers identified the web pages where the structure of the website (the intention of the developers) did not reflect the real behaviour of the visitors. We created the chart of residues for the outliers' identification (Fig. 4). Before that we calculated the basic descriptive characteristics of the variable *Residual* (Table 2).

Outliers Identification Using the Rule ± 2σ. According to the residual theory, we could identify the outliers using the standard rule ± 2σ. It means that we considered the cases that were out of the interval.

Average of differences ± 2 standard deviation of differences.

We calculated the following boundary values for the selected web pages of the website: −5.758532; 5.371501. The chart (Fig. 4) visualizes identified outliers. As we can see, we identified four "suspicious" web pages.

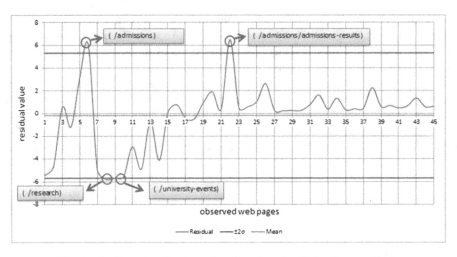

Fig. 4. Outliers visualization using the rule ± 2σ (Color figure online).

Table 2. Descriptive statistics of the variable *Residual*.

	Residual
Valid N	45
Mean	−0,19352
Median	0,431487
Minimum	−5,8136
Maximum	6,485153
LowerQ	−0,46931
UpperQ	0,777322
Range	12,29875
QuartileR	1,246631
QIII + 1,5Q	2,647268
Q1-1,5Q	−2,33926
Std.Dev.	2,782508

We identified two web pages with the residuum greater than +2σ, which were underestimated by the web developers. Even though, the web pages had a few references from other web pages, the visit rate of these web pages was high. It could have been caused by the seasonal importance of the web pages' content. On the other hand, it is clear that these web pages should have references from more relevant web pages on the web site.

At the same time, we identified two "suspicious" web pages with the residuum greater than −2σ. These web pages were overestimated by the website developers. Even though, the web pages had many references from other web pages, these web pages obtained a small visit rate.

Outliers Identification Using Quartiles. We identified the suspicious web pages on the basis of the rule $\pm 2\sigma$ in the previous paragraphs. We analysed 45 web pages and found four suspicious pages. We could ask ourselves if the number of the suspicious web pages was sufficient.

Therefore, we decided to apply the second option for identification of outliers, which is represented by quartiles. In this case, the boundary values are given by the interval $(Q_I - 1,5Q; Q_{III} + 1,5Q)$, where Q_I and Q_{III} represent a lower quartile and an upper quartile, Q is a quartile range. Subsequently, the boundary values -2.339255; 2.647268 were calculated for selected web pages of the website using the quartile range. The chart (Fig. 5) visualises the identified outliers.

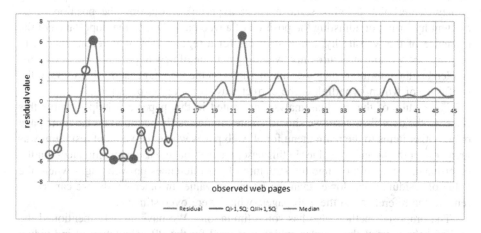

Fig. 5. Outliers visualization using quartiles (Color figure online).

We identified eight new suspicious pages in comparison with the previous method of standard deviation. The expectations of the web page developers were greater than the observed visit rates of the web pages in 7 cases:

- (/university-structure/students-dormitories)
- (/about-the-university)
- (/continuing-education)
- (/media-ukf)
- (/information)
- (/component/user/login)
- (/admissions/registration-to-study/191-registration-instructions)

The developers' expectations were overestimated. These pages had high PR, but low visit rate, i.e. the value of residuum was smaller than $Q_I - 1,5Q$.

In addition, we identified one web page (/about-the-university/contacts-list) with the value of residuum greater than $Q_{III} + 1,5Q$. The expected visit rate was underestimated by the web developers in this case.

The low visit rate in the case of the first and the last identified web pages was caused by the seasonal character of provided information. The content of these two web pages did not attractive for the visitors in the period selected for the experiment.

Other identified suspicious web pages confirmed the previously mentioned problem with the items of the main menu of the website.

5   Discussion

We mentioned in the related work section that a website optimization and restructuring belong to the most frequent application domain of the combination of WUM and WSM methods. There are some common ways to optimize a website structure: adding a group of new hyperlinks, deleting some existent hyperlinks, adjusting the position of an existent hyperlink or adjusting the position of a web page. Due to the importance of the existence of the website hyperlink structure for a regular user, it is recommended to optimize the website structure only by adding a group of new hyperlinks. However, this can lead to the enormous increase in hyperlinks on the web page, and bring users to the confusion of which hyperlink is the correct way to his target page [56].

The results of the residual analysis presented in this paper can help to solve this issue. In the described case, we should create references (hyperlinks) on the homepage or add other items to the main menu for all web pages where the value of residuum was greater than the boundary value. In this case, the web page was underestimated. On the other hand, we should change the structure of the references on web pages where the value of residuum was lower than the boundary value. In other words, we can safely remove the references to the web pages, which were overestimated.

The proposed method also has some limitations. We have already mentioned the limited behaviour of the crawler, which was used for the PR calculation of individual web pages. It is necessary to take into account also other characteristics of the hyperlinks to improve the obtained results. For instance, we have to consider their actual positioning within the layout of each web page which has a strong influence on its foreseen and actual probability of being accessed by a visitor.

Application of another ranking algorithm represents another way how to improve the method described in this paper. For instance, we can use HITS or SALSA algorithms which take into account the content of each web page [56].

Web usage mining methods focus on the behaviour of visitors on the website. The pre-processed log file also contained records about the unwanted clicks. It would also be interesting to detect the unwanted clicks observed for example in the case of website actualities or unwanted pop-ups and remove them from the log file before further analysis.

In general, WUM methods often resulted in a set of rules. We could have evaluated each useful rule subjectively if the rule were really in accordance with the idea of the website developers.

Unfortunately, the usage data used in this case study was limited to the sole fact of accessing a page. The log data contain a wealth data which enable us to assess the importance of a web page. We could extract e.g. the time spend on reading a page and considered it in the process of the web page importance calculation.

We could also identify web pages loaded directly into a browser (without following the link structure of the site). In this case, the users bothered to memorize their addresses and bookmarked them what is a clear indication of their usefulness. We analysed only the internal links of the website using PR algorithm. It could be interesting to include also external links, which target to the website into the PR calculation. These cases were not considered in the presented case study.

The website target group of users has not been taken into account in this case study. If the website provides information for several target groups, we recommend adding this information to the log file in the pre-processing phase. Consequently, we will be able to discover underestimated or overestimated web pages according to the target group and will restructure the website according their needs.

We discovered several underestimated and overestimated web pages in the case study of the university website. Seasonal character of the data, which was used, can limit the final decision about the importance of a particular web page. For this reason, the proposed changes in the structure of the website should be made after sufficiently long period. For instance, this period could take one academic year in the case of the university website.

6 Conclusions

We paid attention to the alternative web structure mining method in this paper and its combination with the web usage mining methods. We chose the algorithm for PR calculation for estimations of importance and quality of the individual web page. The quality of a given web page depends on the number and quality of the web pages that refer to it. We selected the PR method because this method expresses the probability of visiting of given web page by a random visitor.

We presented the case study of the proposed method of identification "suspicious" web pages in the last chapter. Following the conclusions of the previous experiments, we assumed that the expected visit rate correlated with the real visit-rate.

We utilized the potential advantages of joining web structure and web usage mining methods in the residual analysis. We tried to identify the potential problems with the structure of the website. While the sequence rules analysis can only uncover the potential problem of web pages with higher visit rate, the proposed method of residual analysis can also detect other web pages with a smaller visit rate.

In the future research, we will improve the presented method with the aim to eliminate the limitations of this case study, which were mentioned in the discussion. Future work should also focus on the integration of the proposed method with some self-organizing algorithms or other approaches effectively used in the adaptive website design and development.

Acknowledgements. This paper is published with the financial support of the project of Scientific Grant Agency (VEGA), project number VEGA 1/0392/13.

References

1. Srivastava, J., Cooley, R., Deshpande, M., Tan, P.-N.: Web usage mining: discovery and applications of usage patterns from web data. SIGKDD Explor. Newsl. **1**, 12–23 (2000)
2. Liu, Y., Zhang, M., Cen, R., Ru, L., Ma, S.: Data cleansing for web information retrieval using query independent features. J. Am. Soc. Inform. Sci. Technol. **58**, 1884–1898 (2007)
3. Chau, M., Chen, H.: A machine learning approach to web page filtering using content and structure analysis. Decis. Support Syst. **44**, 482–494 (2008)
4. Jacob, A., Olivier, C., Carlos, C.: WITCH: a new approach to web spam detection. Yahoo! Research report no. YR-2008-001 (2008)
5. Castillo, C., Donato, D., Gionis, A., Murdock, V., Silvestri, F.: Know your neighbors: web spam detection using the web topology. In: Conference Know Your Neighbors: Web Spam Detection Using the Web Topology, pp. 423–430. ACM (2006)
6. Gan, Q., Suel, T.: Improving web spam classifiers using link structure. In: Conference Improving Web Spam Classifiers Using Link Structure, pp. 17–20. ACM (2007)
7. Ntoulas, A., Najork, M., Manasse, M., Fetterly, D.: Detecting spam web pages through content analysis. In: Conference Detecting Spam Web Pages Through Content Analysis, pp. 83–92 (2006)
8. Stencl, M., St'astny, J.: Neural network learning algorithms comparison on numerical prediction of real data. In: Matousek, R. (ed.) 16th International Conference on Soft Computing Mendel 2010, pp. 280–285 (2010)
9. Lorentzen, D.G.: Webometrics benefitting from web mining? an investigation of methods and applications of two research fields. Scientometrics **99**, 409–445 (2014)
10. Lili, Y., Yingbin, W., Zhanji, G., Yizhuo, C.: Research on PageRank and hyperlink-induced topic search in web structure mining. In: Conference Research on PageRank and Hyperlink-Induced Topic Search in Web Structure Mining, pp. 1–4 (2011)
11. Wu, G., Wei, Y.: Arnoldi versus GMRES for computing pageRank: a theoretical contribution to google's pageRank problem. ACM Trans. Inf. Syst. **28**, 1–28 (2010)
12. Jain, A., Sharma, R., Dixit, G., Tomar, V.: Page ranking algorithms in web mining, limitations of existing methods and a new method for indexing web pages. In: Proceedings of the 2013 International Conference on Communication Systems and Network Technologies, pp. 640–645. IEEE Computer Society (2013)
13. Ahmadi-Abkenari, F., Selamat, A.: A clickstream based web page importance metric for customized search engines. In: Nguyen, N.T. (ed.) Transactions on Computational Collective Intelligence XII. LNCS, vol. 8240, pp. 21–41. Springer, Heidelberg (2013)
14. Agichtein, E., Brill, E., Dumais, S.: Improving web search ranking by incorporating user behavior information. In: Proceedings of the 29th Annual International ACM SIGIR Conference on Research and Development in Information Retrieval, pp. 19–26. ACM, Seattle (2006)
15. Meiss, M.R., Menczer, F., Fortunato, S., Flammini, A., Vespignani, A.: Ranking web sites with real user traffic. In: Proceedings of the 2008 International Conference on Web Search and Data Mining, pp. 65–76. ACM, Palo Alto (2008)
16. Su, J.-H., Wang, B.-W., Tseng, V.S.: Effective ranking and recommendation on web page retrieval by integrating association mining and PageRank. In: Proceedings of the 2008 IEEE/WIC/ACM International Conference on Web Intelligence and Intelligent Agent Technology, vol. 03, pp. 455–458. IEEE Computer Society (2008)
17. Pabarskaite, Z., Raudys, A.: A process of knowledge discovery from web log data: systematization and critical review. J. Intell. Inf. Syst. **28**, 79–104 (2007)

18. Shutong, C., Congfu, X., Hongwei, D.: Website structure optimization technology based on customer interest clustering algorithm. In: Conference Website Structure Optimization Technology Based on Customer Interest Clustering Algorithm, pp. 802–804 (2008)
19. Wen-long, L., Ye-zheng, L.: A novel website structure optimization model for more effective web navigation. In: Conference A Novel Website Structure Optimization Model for More Effective Web Navigation, pp. 36–41 (2008)
20. Jeffrey, J., Karski, P., Lohrmann, B., Kianmehr, K., Alhajj, R.: Optimizing web structures using web mining techniques. In: Yin, H., Tino, P., Corchado, E., Byrne, W., Yao, X. (eds.) IDEAL 2007. LNCS, vol. 4881, pp. 653–662. Springer, Heidelberg (2007)
21. Wang, H., Liu, X.: Adaptive site design based on web mining and topology. In: Conference Adaptive Site Design Based on Web Mining and Topology, pp. 184–189 (2009)
22. Romero, C., Ventura, S., Zafra, A., Bra, P.D.: Applying web usage mining for personalizing hyperlinks in web-based adaptive educational systems. Comput. Educ. **53**, 828–840 (2009)
23. Park, S., Suresh, N.C., Jeong, B.-K.: Sequence-based clustering for web usage mining: a new experimental framework and ANN-enhanced K-means algorithm. Data Knowl. Eng. **65**, 512–543 (2008)
24. Hay, B., Wets, G., Vanhoof, K.: Web usage mining by means of multidimensional sequence alignment methods. In: Zaïane, O.R., Srivastava, J., Spiliopoulou, M., Masand, B. (eds.) WebKDD 2003. LNCS (LNAI), vol. 2703, pp. 50–65. Springer, Heidelberg (2003)
25. Hay, B., Wets, G., Vanhoof, K.: Segmentation of visiting patterns on web sites using a sequence alignment method. J. Retail. Consum. Serv. **10**, 145–153 (2003)
26. Masseglia, F., Tanasa, D., Trousse, B.: Web usage mining: sequential pattern extraction with a very low support. In: Yu, J.X., Lin, X., Lu, H., Zhang, Y. (eds.) APWeb 2004. LNCS, vol. 3007, pp. 513–522. Springer, Heidelberg (2004)
27. Oyanagi, S., Kubota, K., Nakase, A.: Mining WWW access sequence by matrix clustering. In: Zaïane, O.R., Srivastava, J., Spiliopoulou, M., Masand, B. (eds.) WebKDD 2003. LNCS (LNAI), vol. 2703, pp. 119–136. Springer, Heidelberg (2003)
28. Cooley, R., Mobasher, B., Srivastava, J.: Data preparation for mining world wide web browsing patterns. Knowl. Inf. Syst. **1**(1), 5–32 (1999)
29. Spiliopoulou, M., Faulstich, L.C.: WUM: a tool for web utilization analysis. In: Atzeni, P., Mendelzon, A.O., Mecca, G. (eds.) WebDB 1998. LNCS, vol. 1590, pp. 184–203. Springer, Heidelberg (1999)
30. Chen, M.-S., Park, J.S., Yu, P.S.: Data mining for path traversal patterns in a web environment. In: Conference Data Mining for Path Traversal Patterns in a Web Environment, pp. 385–392 (1996)
31. Berendt, B., Spiliopoulou, M.: Analysis of navigation behaviour in web sites integrating multiple information systems. VLDB J. **9**, 56–75 (2000)
32. Guerbas, A., Addam, O., Zaarour, O., Nagi, M., Elhajj, A., Ridley, M., Alhajj, R.: Effective web log mining and online navigational pattern prediction. Knowl.-Based Syst. **49**, 50–62 (2013)
33. Cooley, R.: Web usage mining: discovery and application of interesting patterns from web data. Ph.D. thesis. University of Minnesota (2000)
34. Schmitt, E., Manning, H., Paul, Y., Tong, J.: Measuring Web Success. Forrester report (1999)
35. Downey, D., Dumais, S., Horvitz, E.: Models of searching and browsing: languages, studies, and applications. In: Proceedings of the 20th International Joint Conference on Artifical Intelligence, pp. 2740–2747. Morgan Kaufmann Publishers Inc., Hyderabad (2007)
36. Chien, S., Immorlica, N.: Semantic similarity between search engine queries using temporal correlation. In: Proceedings of the 14th International Conference on World Wide Web, pp. 2–11. ACM, Chiba (2005)
37. He, D., Göker, A.: Detecting session boundaries from web user logs. In: Conference Detecting Session Boundaries from Web User Logs, pp. 57–66 (2000)

38. Radlinski, F., Joachims, T.: Query chains: learning to rank from implicit feedback. In: Proceedings of the Eleventh ACM SIGKDD International Conference on Knowledge Discovery in Data Mining, pp. 239–248. ACM, Chicago (2005)
39. Huynh, T., Miller, J.: Empirical observations on the session timeout threshold. Inf. Process. Manage. **45**, 513–528 (2009)
40. Zhang, J., Ghorbani, A.A.: The reconstruction of user sessions from a server log using improved time-oriented heuristics. In: Conference The reconstruction of User Sessions from a Server Log Using Improved Time-Oriented Heuristics, pp. 315–322 (2009)
41. Seco, N., Cardoso, N.: Detecting user sessions in the Tumba! query log. Technical report., Faculdade de Ciências da Universidade de Lisboa (2006)
42. Spiliopoulou, M., Mobasher, B., Berendt, B., Nakagawa, M.: A framework for the evaluation of session reconstruction heuristics in web-usage analysis. INFORMS J. Comput. **15**, 171–190 (2003)
43. Gong, W., Baohui, T.: A new path filling method on data preprocessing in web mining. In: Conference A New Path Filling Method on Data Preprocessing in Web Mining, pp. 1033–1035 (2008)
44. Dhawan, S., Lathwal, M.: Study of preprocessing methods in web server logs. Int. J. Adv. Res. Comput. Sci. Softw. Eng. **3**, 430–433 (2013)
45. Li, Y., Feng, B., Mao, Q.: Research on path completion technique in web usage mining. In: Proceedings of the 2008 International Symposium on Computer Science and Computational Technology, vol. 01, pp. 554–559. IEEE Computer Society (2008)
46. Tauscher, L., Greenberg, S.: Revisitation patterns in World Wide Web navigation. In: Proceedings of the ACM SIGCHI Conference on Human Factors in Computing Systems, pp. 399–406. ACM, Atlanta (1997)
47. Chitraa, V., Davamani, A.S.: An Efficient path completion technique for web log mining. In IEEE International Conference on Computational Intelligence and Computing Research (2010)
48. Zhang, C., Zhuang, L.: New path filling method on data preprocessing in web mining. Proc. Comput. Inf. Sci. **1**, 112–115 (2008)
49. Liu, B.: Web data mining. Springer, New York (2007)
50. Brin, S., Page, L.: The anatomy of a large-scale hypertextual web search engine. Comput. Netw. **30**, 107–117 (1998)
51. Page, L., Brin, S., Motwani, R., Winograd, T.: The PageRank citation ranking: bringing order to the web. Technical report, Standford Digital (1998)
52. Pirolli, P., Pitkow, J., Rao, R.: Silk from a sow's ear: extracting usable structures from the web. In: Proceedings of the SIGCHI Conference on Human Factors in Computing Systems, pp. 118–125. ACM, Vancouver (1996)
53. Munk, M., Kapusta, J., Švec, P.: Data preprocessing evaluation for web log mining: reconstruction of activities of a web visitor. Procedia Comput. Sci. **1**, 2273–2280 (2010)
54. Kapusta, J., Munk, M.: Web usage mining: analysis of expeced and observed visit rate UKF (2014)
55. Pilkova, A., Volna, J., Papula, J., Holienka, M.: The influence of intellectual capital on firm performance among slovak SMEs. In: Proceedings of the 10th International Conference on Intellectual Capital, Knowledge Management and Organisational Learning (Icickm-2013), pp. 329–338 (2013)
56. Kumar, P.R., Singh, A.K., Mohan, A.: Efficient methodologies to optimize website for link structure based search engines. In: Conference Efficient Methodologies to Optimize Website for Link Structure Based Search Engines, pp. 719–724 (2013)

Massive Classification with Support Vector Machines

Thanh Nghi Do[1] and Hoai An Le Thi[2($\boxtimes$)]

[1] College of Information Technology, Can Tho University, No 1, Ly Tu Trong Street,
Can Tho 92100, Ninh Kieu District, Vietnam
dtnghi@cit.ctu.edu.vn

[2] Laboratory of Theoretical and Applied Computer Science, University of Lorraine,
Ile de Saulcy, 57045 Metz, France
hoai-an.le-thi@univ-lorraine.fr

Abstract. The new boosting of Least-Squares SVM (LS-SVM), Proximal SVM (PSVM), Newton SVM (NSVM) algorithms aim at classifying very large datasets on standard personal computers (PCs). We extend the PSVM, LS-SVM and NSVM in several ways to efficiently classify large datasets. We developed a row incremental version for datasets with billions of data points. By adding a Tikhonov regularization term and using the Sherman-Morrison-Woodbury formula, we developed new algorihms to process datasets with a small number of data points but very high dimensionality. Finally, by applying boosting including AdaBoost and Arcx4 to these algorithms, we developed classification algorithms for massive, very-high-dimensional datasets. Numerical test results on large datasets from the UCI repository showed that our algorithms are often significantly faster and/or more accurate than state-of-the-art algorithms LibSVM, CB-SVM, SVM-perf and LIBLINEAR.

Keywords: Support vector machine (SVM) · Least-Squares SVM · Proximal SVM · Newton SVM · Boosting · Massive classification

1 Introduction

Since Support Vector Machine (SVM) learning algorithms were first proposed by Vapnik [1], they have been shown to build accurate models with practical relevance for classification, regression and novelty detection. Successful applications of SVMs have been reported for such varied fields as facial recognition, text categorization and bioinformatics [2]. In particular, SVMs using the idea of kernel substitution have been shown to build good models, and they have become increasingly popular classification tools.

However, in spite of their desirable properties, current SVMs cannot easily deal with very large datasets. A standard SVM algorithm requires solving a quadratic or linear programming; so its computational cost is at least $O(m^2)$, where m is the number of training datapoints. Also, the memory requirements of SVM frequently make it intractable. In recent years, the size of data stored

© Springer-Verlag Berlin Heidelberg 2015
N.T. Nguyen (Ed.): Transactions on CCI XVIII, LNCS 9240, pp. 147–165, 2015.
DOI: 10.1007/978-3-662-48145-5_8

in the world doubles every nine months [3,4]. There is a need to scale up these learning algorithms to handle massive datasets. Effective heuristic methods to improve SVM learning time divide the original quadratic programming into series of small problems [5–7]. Incremental learning methods [8–15] improve memory performance for massive datasets by updating solutions in a growing training set without needing to load the entire dataset into memory at once. Parallel and distributed algorithms [11,13] improve learning performance for large datasets by dividing the problem into components that execute on large numbers of net-worked PCs. Active learning algorithms [16,17] choose interesting datapoint sub-sets (active sets) to construct models, instead of using the whole dataset.

In this paper, we describe methods to build boosting of LS-SVM [18], PSVM [19], NSVM [20] algorithms for classifying very large datasets on standard personal computers (PCs), for example, Pentium IV, 1 GB RAM. The LS-SVM classifiers proposed by Suykens and Vandewalle [18] replaces standard SVM optimization inequality constraints with equalities; so the training task only requires solving a system of linear equations instead of a quadratic programming. This makes training times very short. We have extended LS-SVM in three ways.

1. We developed a row-incremental algorithm for classifying massive datasets (billions of points) of dimensionality up to 10^4.
2. Using a Tikhonov regularization term [21] and the Sherman-Morrison-Woodbury formula [22,23], we developed a column-incremental LS-SVM algorithm for very-high-dimensional datasets with small training datapoints, such as bioinformatics microarrays.
3. Applying boosting techniques like AdaBoost (Freund and Schapire [24]) and Arc-x4 (Breiman [25]) to these incremental LS-SVM algorithms, we developed efficient classifiers for massive, very-high-dimensional datasets.

We also applied these ideas to build boosting of other efficient SVM algorithms proposed by Mangasarian and colleagues: Lagrangian SVM [26], Proximal SVM [19] and Newton SVM [20] in the same way, because they have similar properties to LS-SVM. Boosting based on these algorithms is interesting and useful for classification on very large datasets.

Some performances in terms of learning time and accuracy are evaluated on large datasets from the UCI repository [27], Reuters-21578 [28] and RCV1-binary [29] datasets. The results showed that our boosting algorithms are usually much faster and/or more accurate for classification tasks compared with the highly efficient standard SVM algorithm LibSVM (Chang and Lin [29]) and with three recent algorithms, LIBLINEAR (Fan et al. [30]), SVM-perf [31] and CB-SVM (Yu et al. [32]). An example of the effectiveness of the new algorithms is their performance on the 1999 KDD cup dataset. They performed a binary classification of 5 million items from a 41-dimensional input space within 3 min on a standard PC (Pentium 2.4 GHz, 1 GB RAM, Linux).

The remainder of this paper is organized as follows. Section 2 introduces LS-SVM, PSVM and NSVM classifiers. Section 3 describes how to extend LS-SVM, PSVM, NSVM learning algorithms for classifying large datasets. Section 4 explains our procedures for boosting of LS-SVM, PSVM, NSVM. Section 5 presents evaluation results, before the conclusions and future work.

Some notations are used in this paper. All vectors are column vectors unless transposed to row vector by a T superscript. The inner dot product of two vectors, x, y is denoted by $x.y$. The 2-norm of the vector x is denoted by $\|x\|$. The $m \times n$ matrix A is m datapoints in the n-dimensional real space R^n. The classes -1, +1 of m datapoints are denoted by the $m \times m$ diagonal matrix D of -1, +1. e is the column vector of 1. w, b are the normal vector and the scalar of the hyperplane. z is the slack variable and c is a positive constant. I denotes the identity matrix.

2   Least-Squares SVM, Proximal SVM and Newton SVM

2.1   SVM Algorithm

Let us consider a linear binary classification task, as depicted in Fig. 1, with m datapoints x_i ($i = 1, \ldots, m$) in the n-dimensional input space R^n. It is represented by the $m \times n$ matrix A, having corresponding labels $y_i = \pm 1$, denoted by the $m \times m$ diagonal matrix D of ± 1 (where $D_{i,i} = 1$ if x_i is in class +1 and $D_{i,i} = -1$ if x_i is in class -1). For this problem, the SVM algorithms [1] try to find the best separating plane (denoted by the normal vector $w \in R^n$ and the scalar $b \in R^1$), i.e. furthest from both class +1 and class -1. It can simply maximize the distance or margin between the supporting planes for each class ($x.w - b = +1$ for class +1, $x.w - b = -1$ for class -1). The margin between these supporting planes is $2/\|w\|$ (where $\|w\|$ is the 2-norm of the vector w). Any point falling on the wrong side of its supporting plane is considered to be an error. Therefore, the SVM has to simultaneously maximize the margin and minimize the error. The standard SVMs pursue these goals with the quadratic programming of (1).

$$\min \ \Psi(w, b, z) = (1/2)\|w\|^2 + cz \qquad (1)$$
$$s.t. : D(Aw - eb) + z \geq e$$

where $e \in R^m$ denotes the column vector of 1, $z \in R^m$ is the non negative slack vector and the positive constant $c \in R^1$ are used to tune errors, margin size, respectively.

The plane (w, b) is obtained by solving the quadratic programming (1). Then, the classification function of a new datapoint x based on the plane is:

$$predict(x) = sign(w.x - b) \qquad (2)$$

SVM can use some other classification functions, for example a polynomial function of degree d, a RBF (Radial Basis Function) or a sigmoid function. To change from a linear to non-linear classifier, one must only substitute a kernel evaluation in (1) instead of the original dot product. More details about SVM and others kernel-based learning methods can be found in (Cristianini and Shawe-Taylor [33]).

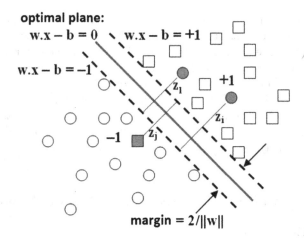

Fig. 1. Linear separation of the datapoints into two classes

Unfortunately, the computational cost requirements of the SVM solutions in (1) are at least $O(m^2)$, where m is the number of training datapoints, making classical SVM intractable for large datasets.

2.2   Least-Squares SVM

The Least-Squares SVM (LS-SVM) proposed by Suykens and Vandewalle [18] has used the equality instead of the inequality constraints in the optimization problem (1) and a least squares 2-norm error in the objective function as follows:

– minimizing the errors by $min\ (c/2)\|z\|^2$
– using the equality constraints $D(Aw - eb) + z = e$

Substituting for z from the constraint in terms of w and b into the objective function Ψ of the quadratic programming (1) yields an unconstrained problem (3):

$$\min\ \Psi(w,b) = (1/2)\|w\|^2 + (c/2)\|e - D(Aw - eb)\|^2 \qquad (3)$$

In the optimal configuration for (3), the gradient with respect to w and b should be zero. This leads the linear equation system of $(n+1)$ variables $(w_1, w_2, \ldots, w_n, b)$ as follows:

$$(w_1,\ w_2,\ \ldots,\ w_n,\ b)^T = (\frac{1}{c}I' + E^T E)^{-1} E^T De \qquad (4)$$

where $E = [A\ -e]$, I' denotes the $(n+1) \times (n+1)$ diagonal matrix whose $(n+1)^{th}$ diagonal entry is zero and the other diagonal entries are 1.

Algorithm 1. Linear LS-SVM algorithm

---

  input :

  Training dataset represented by A and D matrices

  Constant $c > 0$ for tuning errors and margin size

  output:

  (w, b)

  Training:

  begin

  | 1. Create matrix $E = [A \quad -e]$

  | 2. Solve linear equation system (4)

  | 3. Optimal plane (w, b): $(w_1, w_2, \ldots, w_n, b)$

  end

2.3   Proximal SVM

Recent developments for massive linear SVM algorithms proposed by Mangasarian and colleagues reformulate the classification as an unconstrained optimization.

The Proximal SVM (PSVM) proposed by [19] modifies the QP (1) in the same way as LS-SVM. The PSVM also uses the equality constraints in the optimization problem (1) and a least squares 2-norm error in the objective function. Furthermore, the PSVM maximises the margin by $min(1/2)\| w, b \|^2$. By substituting for z from the constraint in terms of w and b into the objective function Ψ leads an unconstrained problem (5):

$$\min \Psi(w, b) = (1/2)\|w, b\|^2 + (c/2)\|e - D(Aw - eb)\|^2 \qquad (5)$$

Setting the gradient with respect to w and b to zero (due to the optimal configuration for (5)) yields the linear equation system of $(n + 1)$ variables $(w_1, w_2, \ldots, w_n, b)$ as follows:

$$(w_1, w_2, \ldots, w_n, b)^T = (\frac{1}{c}I + E^T E)^{-1} E^T De \qquad (6)$$

where $E = [A \quad -e]$, I denotes the identity matrix of size $(n + 1) \times (n + 1)$.

2.4   Newton SVM

The Newton-SVM (NSVM) proposed by [20] reformulates the SVM problem (1) as follows:

- maximizing the margin by $min(1/2)\| w, b \|^2$
- minimizing the errors by $min(c/2)\|z\|^2$

Algorithm 2. Linear PSVM algorithm

---

    input :

        Training dataset represented by A and D matrices

        Constant $c > 0$ for tuning errors and margin size

    output:

        (w, b)

    Training:

    begin

        1. Create matrix $E = [A \quad - e]$

        2. Solve linear equation system (6)

        3. Optimal plane (w, b): $(w_1, w_2, \ldots, w_n, b)$

    end

Substituting for $z = (e - D(Aw - eb))_+$ (where $(x)_+$ replaces negative components of a vector x by zeros) into the objective function Ψ of the quadratic programming (1) yields an unconstrained problem (7):

$$min \ \Psi(w, b) = (c/2)\| \ (e - D(Aw - eb))_+ \ \|^2 + (1/2)\| \ w, b \ \|^2 \qquad (7)$$

By setting $[w_1, w_2, w_n, b]^T$ to u and $[A \quad - e]$ to E, then the unconstrained problem (7) is rewritten by (8):

$$min \ \Psi(u) = (c/2)\| \ (e - DEu)_+ \ \|^2 + (1/2)u^T u \qquad (8)$$

Mangasarian [20] has proposed the finite stepless Newton method for solving the strongly convex unconstrained minimization problem (8).

Mangasarian has proved that the sequence u_i of the Algorithm 3 terminates at the global minimum solution. In most of the tested cases, the stepless Newton algorithm has given the good solution with a number of iterations varying between 5 and 8.

The LS-SVM, PSVM and NSVM algorithms require thus only solutions of linear equations (4, 6 and 11 respectively) of $(n+1)$ variables $(w_1, w_2, \ldots, w_n, b)$ instead of the quadratic programming (1). If the dimensional input space is small enough (less than 10^4), even if there are millions datapoints, these algorithms are able to classify them in some minutes on a PC. The numerical test results [18,19] have shown that they give test correctness compared to standard SVM like LibSVM but their learning tasks are much faster than standard SVMs. An example of its effectiveness [11] is given with the linear classification into two classes of one million datapoints in 20-dimensional input space in 13 s on a PC (2.4 GHz Pentium IV, 512 MB RAM).

Three algorithms can deal with non-linear classification tasks. The training dataset represented by the $m \times n$ matrix A at the input is replaced by the $m \times m$ kernel matrix K, where K is a non linear kernel matrix created by whole dataset A and the support vectors being A too, e.g.:

Algorithm 3. Linear Newton SVM algorithm

---

input :

  Training dataset represented by A and D matrices

  Constant $c > 0$ for tuning errors and margin size

output:

  (w, b)

Training:

begin

  1. Create matrix $E = [A \quad -e]$

  2. Starting with $u_0 \in R^{n+1}$ and $i = 0$

  3. **repeat**

    3.1. The gradient of Ψ at u_i is:

$$\triangledown \Psi(u_i) = c(-DE)^T(e - DEu_i)_+ + u_i \qquad (9)$$

    3.2. The generalized Hessian of Ψ at u_i is:

$$\partial^2 \Psi(u_i) = c(-DE)^T diag([e - DEu_i]_*)(-DE) + I \qquad (10)$$

    with $diag([e - DEu_i]_*)$ denotes the $(n+1) \times (n+1)$ diagonal matrix whose j^{th} diagonal entry is sub-gradient of the step function $(e - DEu_i)_+$

    and I is the identity matrix of size $(n+1) \times (n+1)$.

    3.3. Updating

$$u_{i+1} = u_i - \partial^2 \Psi(u_i)^{-1} \triangledown \Psi(u_i) \qquad (11)$$

    3.4. Increment $i = i + 1$

  until $\triangledown\Psi(u_i) < tol$;

  4. Optimal plane (w, b): $(w_1, w_2, \ldots, w_n, b)$ via u_i

end

- A degree d polynomial kernel of two datapoints x_i, x_j: $K_{i,j} = (x_i.x_j + 1)^d$
- A radial basis kernel of two datapoints x_i, x_j: $K_{i,j} = \exp(-\Upsilon\|x_i - x_j\|^2)$

These algorithms using a $m \times m$ kernel matrix K require very large memory size and execution time.

3 Extensions of LS-SVM, PSVM, NSVM for Large Datasets

Although the LS-SVM, PSVM, NSVM algorithms can efficiently classify large datasets, it still needs to keep the whole dataset in the memory. This is impractical for large datasets: for example, 1 billion 20-dimensional data points would require 80 GB of RAM. Most machine-learning algorithms have problems dealing with the challenge of large datasets. Our investigation aims to scale up the LS-SVM, PSVM, NSVM algorithms for mining very large datasets on standard PCs. First, we addressed the memory issue with incremental learning algorithms. These avoid loading the whole dataset in main memory by considering only subsets of the data at a time updating the solution in a growing training set.

3.1 Classifying Very Large Number of Datapoints with LS-SVM, PSVM and NSVM

Datasets with a very large number m of datapoints but a moderate number n of dimensions can be decomposed into k small blocks of rows A_1, D_1, ..., A_k, D_k. as follows:

$$A = \begin{bmatrix} A_1 \\ \vdots \\ A_k \end{bmatrix}, D = \begin{bmatrix} D_1 & & \\ & \ddots & \\ & & D_k \end{bmatrix}, and\, e = \begin{bmatrix} e_1 \\ \vdots \\ e_k \end{bmatrix}$$

And then,

$$E = [A \quad -e] = \begin{bmatrix} E_1 \\ \vdots \\ E_k \end{bmatrix}$$

Therefore,

$$E^T E = \sum_{i=1}^{k} E_i^T E_i$$

$$E^T De = \sum_{i=1}^{k} E_i^T D_i e_i$$

The row-incremental algorithm of the LS-SVM in (4) can simply incrementally compute the solution of the linear equation system (12):

$$(w_1, w_2, \ldots, w_n, b)^T = (\frac{1}{c}I' + \sum_{i=1}^{k} E_i^T E_i)^{-1} \sum_{i=1}^{k} E_i^T D_i e_i \qquad (12)$$

In the same incermetal way as LS-SVM, the row-incremental algorithm of the PSVM in (6) computes the solution of the linear equation system (13):

$$(w_1, w_2, \ldots, w_n, b)^T = (\frac{1}{c}I + \sum_{i=1}^{k} E_i^T E_i)^{-1} \sum_{i=1}^{k} E_i^T D_i e_i \qquad (13)$$

The row-incremental algorithm of the NSVM incrementally computes the gradient of Ψ at u_i in (9) and the generalized Hessian of Ψ at u_i in (10).

Consequently, the row-incremental LS-SVM, PSVM, NSVM algorithms can handle massive datasets on a PC. The accuracy of the row-incremental algorithm is exactly the same as the original one. If the dimension of the input space is small enough (less than 10^4), even if there are billions of datapoints, the row-incremental algorithms are able to classify them on a simple PC (Pentium IV, 512 MB RAM). The algorithms only need to store a small $(n+1) \times (n+1)$ matrix and two $(n+1)$-components vectors in memory between two successive steps. The numerical test [11] has shown the row-incremental algorithms can classify one billion datapoints in 20-dimensional input into two classes in 26 min and 40 s (excluding approximately one hour of time to read data from disk) on a PC Pentium-IV 2.4 GHz, 512 MB RAM.

Algorithm 4. Row incremental LS-SVM algorithm

---

input :
>   Training dataset represented by k blocks $A_1, D_1, \ldots, A_k, D_k$
>   Constant $c > 0$ for tuning errors and margin size

output:
> (w, b)

Training:
begin
>   1. Initial $(n + 1) \times (n + 1)$ matrix $E^T E = 0$
>   2. Initial $(n + 1) - components$ vector $d = E^T De = 0$
>   3. for $i \leftarrow 1$ to k do
>>   3.1. Load A_i and D_i
>>   3.2. Create matrix $E_i = \begin{bmatrix} A_i & -e \end{bmatrix}$
>>   3.3. Compute $E^T E = E^T E + E_i^T E_i$
>>   3.4. Compute $d_i = E_i^T D_i e$
>>   3.5. Compute $d = d + d_i$
>   end
>   4. Solve linear equation system (12) with $E^T E = \sum_{i=1}^{k} E_i^T E_i$ and
> $d = E^T De = \sum_{i=1}^{k} E_i^T D_i e_i$
>   5. Optimal plane (w, b): $(w_1, \ w_2, \ \ldots, \ w_n, \ b)$
end

3.2   Classifying Very-High-Dimensional Datasets with LS-SVM, PSVM and NSVM

In some applications like bioinformatics, the datasets usually have a very large number n of dimensions and few training datapoints. Thus, the $(n+1) \times (n+1)$ matrix $E^T E$ is too large and the solution of the linear equation system of $(n+1)$ variables (w, b) has a high computational cost. To adapt the LS-SVM, PSVM and NSVM algorithms to these problems, we apply the Sherman-Morrison-Woodbury formula [22, 23] to the linear equation systems (4, 6, 11).

$$(A + UV^T)^{-1} = A^{-1} - A^{-1}U(I + V^T A^{-1}U)^{-1}V^T A^{-1} \qquad (14)$$

For LS-SVM, we added a Tikhonov regularization term [21] in the linear equation system (4), permitting us to apply the Sherman-Morrison-Woodbury formula to LS-SVM as follow.

For a suitably chosen Tikhonov factor $\delta > 0$ added to the linear equation system (4), we obtain a regularized version of LS-SVM (15):

$$(w_1, \ w_2, \ \ldots, \ w_n, \ b)^T = (H + E^T E)^{-1} E^T De \qquad (15)$$

where H denotes the $(n+1) \times (n+1)$ diagonal matrix whose $(n+1)^{th}$ diagonal entry is δ and the other diagonal entries are $(1/c) + \delta$.

Thus, applying the Sherman-Morrison-Woodbury formula (14) to the right side of (15) yields the new linear equation system (16):

$$(H + E^T E)^{-1} E^T D e =$$
$$(H^{-1} - H^{-1} E^T (I + E H^{-1} E^T)^{-1} E H^{-1}) E^T D e =$$
$$(H^{-1} E^T - H^{-1} E^T (I + E H^{-1} E^T)^{-1} E H^{-1} E^T) D e =$$
$$H^{-1} E^T (I - (I + E H^{-1} E^T)^{-1} E H^{-1} E^T) D e$$

And then, we obtain:

$$(w_1,\ w_2,\ \ldots,\ w_n,\ b)^T = H^{-1} E^T (I - (I + E H^{-1} E^T)^{-1} E H^{-1} E^T) D e \quad (16)$$

With PSVM, we apply the Sherman-Morrison-Woodbury formula to the linear equation system (6), the new PSVM formula for very large number of dimensions is:

$$(w_1,\ w_2,\ \ldots,\ w_n,\ b)^T = c E^T (I - (I/c + E E^T)^{-1} E E^T) D e \quad (17)$$

The new LS-SVM, PSVM are accomplished through the linear equation systems (16, 17 respectively) which require inverting the $(m) \times (m)$ matrices $((I + E H^{-1} E^T)$ or $(I/c + E E^T))$. Thus, the computation and storage cost depends on the number m of training datapoints, not the number n of dimensions. These formulations can handle datasets with a very large number of dimensions and few training datapoints.

With NSVM, by setting $Q = diag(\sqrt{c * [e - D E u_i]_*})$ and $P = Q(-DE)$, the inverse matrix $\partial^2 \Psi(u_i)^{-1}$ can be re-written as follow (18) :

$$\partial^2 \Psi(u_i)^{-1} = (I + P^T P)^{-1} \quad (18)$$

Thus, we apply the Sherman-Morrison-Woodbury formula (14) to the right part of (18), we obtain the inverse matrix $\partial^2 f(u_i)^{-1}$ as (19):

$$\Rightarrow \partial^2 \Psi(u_i)^{-1} = I - P^T (I + P P^T)^{-1} P \quad (19)$$

The updating step of u_i in (11) using $\partial^2 \Psi(u_i)^{-1}$ formula in (19) only depends on the inversion of the $(m) \times (m)$ matrix $(I + P P^T)$. And then, the cost of storage and computation of NSVM depends on the number of training data. This new NSVM formulation can handle datasets with very large number of dimensions and few training data.

4   Boosting of LS-SVM, PSVM and NSVM

For classifying massive datasets with simultaneously large number (at least 10^4) of datapoints and dimensions, e.g. text categorization problems, there are at least two problems to solve: the learning time increases dramatically with the training data size and the memory requirement increases according to data size. Although the incremental algorithms of LS-SVM, PSVM and NSVM can in fact add new training data efficiently these algorithms need to store and invert a matrix whose size is the square of the number of datapoints or dimensions. Therefore, they can not avoid requiring too much main memory and very high computational time while processing simultaneously large number datapoints and dimensions.

To scale LS-SVM, PSVM, NSVM to large datasets, we have applied the boosting approach like AdaBoost (Freund and Schapire [24]) and Arc-x4 (Breiman [25]) to the LS-SVM, PSVM, NSVM algorithms. This brings out two advantages. The first one is to be able to solve the scaling problem and the second one is the achievement of the classification accuracy. Furthermore, the boosting approach is consistent in terms of the margins (Reyzin and Schapire [34]). We briefly explain how construct boosting of LS-SVM, PSVM, NSVM. The AdaBoost algorithm introduced by Freund and co-workers is a general method for improving the accuracy of any given weak learning algorithm. The algorithm calls repeatedly a given weak or base learning algorithm t times so that each boosting step concentrates mostly on the errors produced by the previous step. For achieving this goal, it needs to maintain a distribution weights over the training points. Initially, all weights are uniform equals and at each boosting step the weights of incorrectly classified examples are increased so that the weak learner is forced to focus on the hard examples in the training set. The final predict model is a weighted majority vote of t weak classifiers. Alternately, we consider the LS-SVM, PSVM, NSVM algorithms as a weak algorithm because at each boosting step it only learns on a subset of the training set according to the distribution weights over the training examples. Extensions of LS-SVM, PSVM, NSVM described in section (3) can be adapted to solve large sizes of subset. With large number of datapoints in dimensional input space is small enough (less than 10^4), we build AdaBoost based on the row-incremental of LS-SVM, PSVM, NSVM.

For dealing with a very large number of dimensions and few training data-points or simultaneously large number of datapoints and dimensions, we build AdaBoost based on the extensions of LS-SVM, PSVM, NSVM for very-high-dimensional datasets. Note that LS-SVM, PSVM, NSVM outperform SVM standard algorithms in term of the learning time on subsets (less than the original training set). Therefore, we are interested in boosting of LS-SVM, PSVM, NSVM instead of any other standard SVM. Algorithm 5 describes the AdaBoost of LS-SVM, PSVM, NSVM.

We have also applied the Arc-x4 algorithm to LS-SVM in the same way. Breiman has proposed the Arc-x4 algorithm to study the behaviour of AdaBoost. The Arc-x4 algorithm is similar to AdaBoost. It uses un-weighted majority vote instead of using weighted one and it also has a different way for updating the distribution weights over the training points based on the number of misclas-sifications of points by last classifiers. The Arc-x4 performs equally well as the AdaBoost. Thus, we have constructed the Arc-x4 of LS-SVM to mine very large datasets.

Boosting of LS-SVM, PSVM, NSVM algorithms have shown performances concerning the learning time and the classification accuracy. E.g. 5 millions of datapoints in 41-dimensional input are classified into two classes in 3 min on one PC (Pentium 2.4 GHz, 1 GB RAM) or 5 hundred thousands of datapoints

---

Algorithm 5. AdaBoost of LS-SVM, PSVM, NSVM

---

input :

      Training dataset with m datapoints: $\{x_i, y_i\}_{i=1,m}$, $x_i \in R^n$ and $y_i \in \pm 1$

      Constant $c > 0$ for tuning errors and margin size

      Number of boosting iterations $MaxIt$

output:

      (w, b)

Training:

begin

    1. Initial distribution of m datapoints: $p_1(i) = 1/m$

    2. **for** $t \leftarrow 1$ **to** $MaxIt$ **do**

        2.1. Sampling S_i of datapoints using p_t

        2.2. Learning $\{LS, P, N\}SVM_t$ from S_i

        2.3. Computing predicting errors of $\{LS, P, N\}SVM_t$:

        $\epsilon_t = \sum_{i=1}^{m} p_t(i) 1_{\{LS,P,N\}SVM_t(x_i) \neq y_i}$

        2.4 **if** $\epsilon_t > 0.5$ **then** stop

        2.5. Choosing $\alpha_t = \frac{1}{2} ln \frac{1-\epsilon_t}{\epsilon_t}$

        2.6. Computing normalized factor $Z_t = 2\sqrt{\epsilon_t(1-\epsilon_t)}$

        2.7. Updating distribution of m datapoints:

        if ($\{LS, P, N\}SVM_t(x_i) \neq y_i$) **then**

            $p_{t+1}(i) = p_t(i) \frac{exp(\alpha_t)}{Z_t}$

        else

            $p_{t+1}(i) = p_t(i) \frac{exp(-\alpha_t)}{Z_t}$

    end

    3. Optimal plane (w, b) by aggregating models $\{LS, P, N\}SVM_t$ and weights α_t

end

---

in 54-dimensional input are classified into two classes in 30 s. These above ideas are also used to build boosting of efficient algorithms proposed by Mangasarian and his colleagues.

5 Evaluation of Strategies and Comparison with Other SVM Methods

We have implemented our algorithms in C/C++ on PC Linux, we have also used the high performance linear algebra library, Lapack++ (Dongarra et al. [35]). We have also implemented some strategies for performing sparse data matrix computations. Thus, our boosting of LS-SVM, PSVM, NSVM algorithms are able to deal with large datasets in linear classification tasks. All tests were run under Linux on a single 2.4-GHz Pentium-4 PC with 2 GB RAM.

We evaluated the learning time and classification accuracy of our boosting strategies of LS-SVM, PSVM, NSVM algorithms. However, the PSVM is very closed to the LS-SVM. Therefore we only present classification results of four algorithms, including {AdaB, Arcx4}-LS-SVM and {AdaB, Arcx4}-NSVM. We compare their performance in terms of correctness and training time with the LibSVM [29] using a linear kernel, using published results for the more recent SVM-Perf [31] and CB-SVM [32] algorithms where it failed to complete the tasks.

The datasets from small to large size listed in Table 1 were selected from the UCI repository [27], benchmark collections for text categorization, Reuters-21578 (Lewis [28]) and RCV1-binary (Chang and Lin [29]). We have performed some pre-processing steps with the datasets. The first, the nominal attributes from the Adult are converted into binary ones.

Table 1. Dataset description

Datasets	Classes	Datapoints	Dimensions	Evaluation methods
Adult	2	48842	14	32561 Trn - 16281 Tst
Reuters-21578	135	10789	29406	7770 Trn - 3019 Tst
RCV1-binary	2	697641	47236	20242 Trn - 677399 Tst
Forest covertype	8	581012	54	10-fold
KDD cup 1999	5	5209458	41	4898429 Trn - 311029 Tst

Due to the evaluation of performance for datasets with both large numbers of datapoints and dimensions we used the Reuters-21578 text collection and RCV1-binary, both known to be good benchmarks for text categorization. We used the Bow software program (McCallum [36]) to pre-process the Reuters-21578 dataset, representing each document as a vector of words. Without any feature selection, this yielded a dataset of 29406 dimensions (words). As common for datasets with more than two classes, we trained one learner for each class (considered as +1 class), with all the others considered as the -1 class. Table 2 reports the average of Precision and Recall (breakeven point) for the 10 largest categories. Table 3 presents training time. The results showed that our boosting of LS-SVM, NSVM algorithms have comparable performance to LibSVM, typically having better accuracy and but slightly longer training time.

We used the pre-processing of Chang and Lin for the RCV1 dataset, considering CCAT, ECAT as the positive class and GCAT, MCAT as the negative class, and removing instances that occurred in both the positive and negative classes. The classification results for this dataset in Table 4 show that our boosting of LS-SVM algorithms are much faster than LibSVM (18 times) for this dataset, while also outperforming in terms of accuracy (Figs. 2 and 3)

Table 5 also presents the results for three other very large datasets of moderate dimensionality. For the Adult dataset, our algorithms are a factor of 190 faster than LibSVM, with nearly identical accuracy.

For Forest cover type dataset, both of our algorithms classified the 2 largest classes (211840 Spruce-Fir and 283301 Lodgepole-Pine datapoints) using 54

Table 2. Classification performance for ten largest categories of Reuters-21578

	Breakeven accuracy(%)				
	AdaB	Arcx4	AdaB	Arcx4	
	LS-SVM	LS-SVM	NSVM	NSVM	LibSVM
Earn	**98.41**	98.32	98.32	<u>98.36</u>	98.02
Acq	**96.57**	96.22	95.66	<u>96.44</u>	95.66
Money-fx	79.49	<u>80.90</u>	80.36	**81.04**	75.72
Grain	<u>90.75</u>	90.73	**91.07**	90.01	89.33
Crude	**89.83**	88.98	88.89	<u>89.58</u>	86.62
Trade	78.18	76.39	**79.74**	<u>79.05</u>	77.46
Interest	78.32	78.16	<u>78.59</u>	**80.15**	75.57
Ship	84.60	83.72	<u>85.63</u>	**86.25**	83.00
Wheat	86.38	**88.12**	85.74	<u>86.94</u>	85.58
Corn	88.97	**89.12**	88.14	88.05	<u>88.99</u>

Table 3. Training time for ten largest categories of Reuters-21578

	Training time (secs)				
	AdaB	Arcx4	AdaB	Arcx4	
	LS-SVM	LS-SVM	NSVM	NSVM	LibSVM
Earn	<u>8.42</u>	**7.24**	12.22	13.15	9.40
Acq	<u>8.47</u>	**6.85**	5.81	15.74	10.78
Money-fx	<u>7.68</u>	9.30	11.29	26.52	**7.42**
Grain	9.86	**4.74**	12.67	12.99	<u>5.05</u>
Crude	<u>8.05</u>	9.25	16.85	17.38	**5.70**
Trade	8.01	**3.02**	11.81	10.94	<u>5.73</u>
Interest	13.33	<u>11.66</u>	6.02	11.62	**8.69**
Ship	<u>14.21</u>	16.80	16.07	25.13	**5.36**
Wheat	14.10	**2.41**	9.56	10.40	<u>3.54</u>
Corn	<u>6.28</u>	6.62	19.48	13.88	**3.79**

Table 4. Classification performance for massive datasets

	Accuracy(%)				
	AdaB	Arcx4	AdaB	Arcx4	
	LS-SVM	LS-SVM	NSVM	NSVM	LibSVM
Adult	85.08	85.03	85.18	**85.34**	<u>85.29</u>
Forest cover type	<u>77.16</u>	76.69	**77.18**	76.72	N/A
Kdd cup 99	91.96	**92.73**	92.31	<u>92.65</u>	N/A
RCV1-binary	95.45	<u>95.77</u>	95.57	**95.93**	93.28

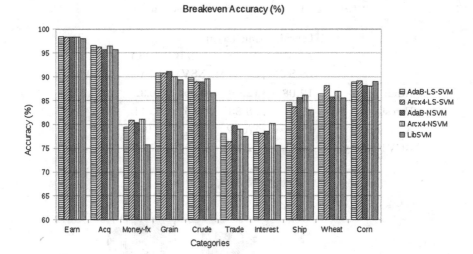

Fig. 2. Classification performance for ten largest categories of Reuters-21578

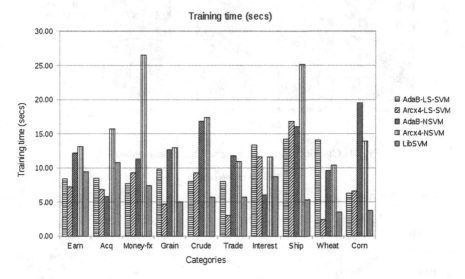

Fig. 3. Training time for ten largest categories of Reuters-21578

attributes in under 30 s. LibSVM ran for 21 days without any result. However, recently-published results indicate that the SVM-perf algorithm (Joachims [31]) performed this classification in 171 s on a 3.6 GHz Intel Xeon processor with 2 GB RAM. Making reasonable adjustments for processor speed, this indicates that our boosting of LS-SVM algorithms are probably about 8 times faster than SVM-Perf.

Table 5. Training time for massive datasets

	Learning time (secs)				
	AdaB	Arcx4	AdaB	Arcx4	
	LS-SVM	LS-SVM	NSVM	NSVM	LibSVM
Adult	**12.08**	**12.08**	21.00	49.32	2472.59
Forest cover type	29.83	**28.83**	<u>29.52</u>	43.14	N/A
Kdd cup 99	191	176.62	<u>172.23</u>	**149.92**	N/A
RCV1-binary	54.10	**42.21**	<u>48.97</u>	178.57	787.14

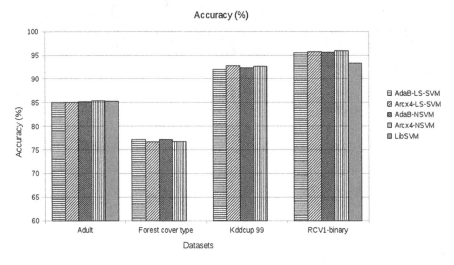

Fig. 4. Classification performance for massive datasets

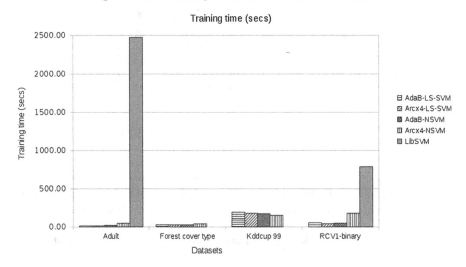

Fig. 5. Training time for massive datasets

The KDD Cup 1999 dataset consists of network data indicating either normal connections (negative class) or attacks (postive class). LibSVM ran out of memory. CB-SVM (Yu et al. [32]) has classified the dataset with over 90 % accuracy in 4750 s on a Pentium 800 MHz with 1 GB RAM, while our algorithms achieved 92 % accuracy in only 180 s. Adjusting for processor speed, they appear to be about a factor of 9 times faster than CB-SVM.

It seems that our Arcx4-NSVM achieves high accuracies with longer training time than {AdaB, Arcx4}-LS-SVM algorithms (Figs. 4 and 5).

We are curious about the recent LIBLINEAR (Fan et al. [30]). It is well-known as recent advances of large-scale linear classification. LIBLINEAR classifies Forest cover type dataset in 137.66 s with 64.29 % accuracy. This result shows that our boosting algorithms are a factor of 4 faster than LIBLINEAR, with a significant improvement 13 % in terms of accuracy. For KDD Cup 1999 dataset, LIBLINEAR achieves 92.15 % accuracy in 1521.07 s. It is about a factor of 8 times slower than our ones.

For concluding, the obtained experimental results show that our boosting of LS-SVM, PSVM, NSVM algorithms are fast and accurate to classify massive datasets having large number of datapoints or large number of dimensions or simultaneously very large number of datapoints and dimensions on PCs.

6 Conclusion and Future Work

This paper presents boosting of LS-SVM, PSVM, NSVM algorithms developed by us that can perform linear classification on very large and high-dimensional datasets on ordinary PCs. The main ideas are building incremental strategies and boosting strategies of the LS-SVM, PSVM, NSVM. The accuracy of the incremental algorithms is exactly the same as that of the original in-memory versions. If the dimensionality of the input space is moderate (less than 10^4), row-incremental LS-SVM,PSVM, NSVM can classify billions of datapoints in reasonable time, even on a simple PC (Pentium IV, 512 MB RAM).

Adding a Tikhonov regularization term and also using the Sherman-Morrison-Woodbury formula enabled us to also build new LS-SVM, PSVM, NSVM algorithms capable of processing datasets of very high dimensionality.

Finally, we have extended this idea by applying boosting approaches like AdaBoost and Arc-x4 to LS-SVM, PSVM, NSVM algorithms. This allowed us to build mining algorithms for massive datasets that are also very-high-dimensional. Numerical test results have shown that our boosting of LS-SVM, PSVM, NSVM algorithms are usually significantly faster and/or more accurate compared with state-of-the-art algorithms such as LibSVM and the recently-proposed algorithms LIBLINEAR, SVM-Perf and CB-SVM.

A forthcoming improvement will be to extend these approaches for multi-class and non-linear classification tasks.

References

1. Vapnik, V.: The Nature of Statistical Learning Theory. Springer, New York (1995)
2. Guyon, I.: Web page on svm applications (1999). http://www.clopinet.com/ isabelle/Projects/-SVM/app-list.html
3. Fayyad, U., Piatetsky-Shapiro, G., Uthurusamy, R.: Summary from the kdd-03 panel - data mining: the next 10 years. SIGKDD Explor. **5**(2), 191–196 (2004)
4. Lyman, P., Varian, H.R., Swearingen, K., Charles, P., Good, N., Jordan, L., Pal, J.: How much information (2003). http://www.sims.berkeley.edu/research/projects/ how-much-info-2003/
5. Boser, B., Guyon, I., Vapnik, V.: An training algorithm for optimal margin classifiers. In: Proceedings of 5th ACM Annual Workshop on Computational Learning Theoryof 5th ACM Annual Workshop on Computational Learning Theory, pp. 144–152. ACM (1992)
6. Osuna, E., Freund, R., Girosi, F.: An improved training algorithm for support vector machines. In: Principe, J., Gile, L., Morgan, N., Wilson, E., (eds.) Neural Networks for Signal Processing VII, pp. 276–285 (1997)
7. Platt, J.: Fast training of support vector machines using sequential minimal optimization. In: Schölkopf, B., Burges, C., Smola, A. (eds.) Advances in Kernel Methods - Support Vector Learning, pp. 185–208. MIT Press, Cambridge (1999)
8. Cauwenberghs, G., Poggio, T.: Incremental and decremental support vector machine learning. Adv. Neural Inf. Process. Syst. **13**, 409–415 (2001)
9. Do, T.N., Poulet, F.: Incremental svm and visualization tools for bio-medical data mining. In: Proceedings of Workshop on Data Mining and Text Mining in Bioinformatics, pp. 14–19 (2003)
10. Do, T.N., Poulet, F.: Towards high dimensional data mining with boosting of psvm and visualization tools. In: Proceedings of 6th International Conference on Entreprise Information Systems, pp. 36–41 (2004)
11. Do, T.N., Poulet, F.: Classifying one billion data with a new distributed svm algorithm. In: Proceedings of 4th IEEE International Conference on Computer Science, Research, Innovation and Vision for the Future, pp. 59–66. IEEE Press (2006)
12. Fung, G., Mangasarian, O.: Incremental support vector machine classification. In: Proceedings of the 2nd SIAM International Conference on Data Mining (2002)
13. Poulet, F., Do, T.N.: Mining very large datasets with support vector machine algorithms. In: Camp, O., Filipe, J., Hammoudi, S., Piattini, M., et al. (eds.) Enterprise Information Systems V, pp. 177–184. Kluwer Academic Publishers, Dordrecht (2004)
14. Do, T.N., Le-Thi, H.A.: Classifying large datasets with svm. In: Proceedings of 4th International Conference on Computational Management Science (2007)
15. Syed, N., Liu, H., Sung, K.: Incremental learning with support vector machines. In: Proceedings of the ACM SIGKDD International Conference on KDD. ACM (1999)
16. Do, T.N., Poulet, F.: Mining very large datasets with svm and visualization. In: Proceedings of 7th International Conference on Entreprise Information Systems, pp. 127–134 (2005)
17. Tong, S., Koller, D.: Support vector machine active learning with applications to text classification. In: Proceedings of the 17th International Conference on Machine Learning, pp. 999–1006. ACM (2000)

18. Suykens, J., Vandewalle, J.: Least squares support vector machines classifiers. Neural Process. Lett. **9**(3), 293–300 (1999)

19. Fung, G., Mangasarian, O.: Proximal support vector classifiers. In: Proceedings of the ACM SIGKDD International Conference on KDD, pp. 77–86. ACM (2001)

20. Mangasarian, O.: A finite newton method for classification problems. Technical report 01–11, Data Mining Institute, Computer Sciences Department, University of Wisconsin (2001)

21. Tikhonov, A.N.: On the stability of inverse problems. Dokl Akad. Nauk SSSR **39**(5), 195–198 (1943)

22. Golub, G., Loan, C.V.: Matrix Computations, 3rd edn. John Hopkins University Press, Baltimore (1996)

23. Rao, C.: Linear Statistical Inference and Its applications. Wiley, New York (1965)

24. Freund, Y., Schapire, R.: A short introduction to boosting. J. Japan. Soc. Artif. Intell. **14**(5), 771–780 (1999)

25. Breiman, L.: Arcing classifiers. Ann. Stat. **26**(3), 801–849 (1998)

26. Mangasarian, O., Musicant, D.: Lagrangian support vector machines. J. Mach. Learn. Res. **1**, 161–177 (2001)

27. Frank, A., Asuncion, A.: UCI machine learning repository (2010). http://www.ics. uci.edu/~mlearn/MLRepository.html

28. Lewis, D.: Reuters-21578 text classification test collection (1997). http://www. david-dlewis.com/resources/testcollections/reuters21578/

29. Chang, C.C., Lin, C.J.: LIBSVM - a library for support vector machines (2001). http://www.csie.ntu.edu.tw/~cjlin/libsvm

30. Fan, R., Chang, K., Hsieh, C., Wang, X., Lin, C.: LIBLINEAR: a library for large linear classification. J. Mach. Learn. Res. **9**(4), 1871–1874 (2008)

31. Joachims, T.: Training linear svms in linear time. In: Proceedings of the ACM SIGKDD International Conference on KDD, pp. 217–226. ACM (2006)

32. Yu, H., Yang, J., Han, J.: Classifying large data sets using svms with hierarchical clusters. In: Proceedings of the ACM SIGKDD International Conference on KDD, pp. 306–315. ACM (2003)

33. Cristianini, N., Shawe-Taylor, J.: An Introduction to Support Vector Machines and Other Kernel-Based Learning Methods. Cambridge University Press, Cambridge (2000)

34. Reyzin, L., Schapire, R.: How boosting the margin can also boost classifier complexity. In: Proceedings of the 23rd International Conference on Machine Learning, pp. 753–760. ACM (2006)

35. Dongarra, J., Pozo, R., Walker, D.: LAPACK++: a design overview of object-oriented extensions for high performance linear algebra. In: Proceedings of Supercomputing, pp. 162–171 (1993)

36. McCallum, A.: Bow: a toolkit for statistical language modeling, text retrieval, classification and clustering (1998). http://www-2.cs.cmu.edu/~mccallum/bow

On a Multi-agent Distributed Asynchronous Intelligence-Sharing and Learning Framework

Shashi Shekhar Jha$^{(\boxtimes)}$ and Shivashankar B. Nair

Department of Computer Science and Engineering,
Indian Institute of Technology Guwahati, Guwahati 781039, India
{j.shashi,sbnair}@iitg.ernet.in

Abstract. The current digital era is flooded with devices having high processing and networking capabilities. Sharing of information, learning and adaptation in such highly distributed systems can greatly enhance their performance and utility. However, achieving the same in the presence of asynchronous entities is a complex affair. Multi-agent system paradigms possess intrinsic similarities with these distributed systems and thus provide a fitting platform to solve the problems within. Traditional approaches to efficient information sharing and learning among autonomous agents in distributed environments incur high communication overheads. Non-conventional tactics based on social insect colonies provide natural solutions for transfer of social information in highly distributed and dense populations. This paper portrays a framework to achieve distributed and asynchronous sharing of intelligence and consequent learning among the entities of a networked distributed system. This framework couples localized communication with the available multi-agent technologies to realize asynchronous intelligence-sharing and learning. The framework takes in a user-defined objective together with a learning algorithm as inputs and facilitates cooperative learning among the agents using the mechanisms embedded within. The proposed framework has been implemented using *Typhon* agent framework over a LAN. The results obtained from the experiments performed using both static and dynamic LANs, substantiate the applicability of the proposed framework in real distributed mobile computing environments.

Keywords: Multi-agent learning · Distributed intelligence · Mobile agents · Typhon · Emulation

1 Introduction

The drastic increase in the number of computational entities or devices in our surroundings has propelled research towards devising decentralized and distributed approaches for solving complex real-world problems. New paradigms, such as Cyber-Physical Systems (CPS) [39] and the Internet-of-Things (IoT) [2], wherein researchers attempt to evolve large-scale intelligent and autonomous applications, require a high degree of co-ordination among the numerous small

© Springer-Verlag Berlin Heidelberg 2015
N.T. Nguyen (Ed.): Transactions on CCI XVIII, LNCS 9240, pp. 166–200, 2015.
DOI: 10.1007/978-3-662-48145-5_9

scale computing units that comprise these systems. The scale of the system and the computational and communication related complexities restrict the use of traditional centralized approaches to achieve such co-ordination. Though these approaches provide a single-point control of the overall system, they are less flexible in terms of reconfigurability and failures. Researchers have thus digressed to conceive non-traditional paradigms [28,41] to develop such distributed and autonomous large scale systems.

One of the approaches to solve problems that lie in distributed environments is the multi-agent system paradigm. Multi-agent systems [12] focus on the collective behaviours of agents and the complexities springing from their interactions. They form a fitting platform to realize truly distributed solutions in their most natural forms. Learning and sharing of information in multi-agent systems is a complex task both conceptually and technically [40] as it involves multiple learners wherein each agent tries to learn and adapt concurrently and in conjunction with the others. Research in the domain of multi-agent learning [1] has largely bifurcated into two streams - those that use reinforcement learning and the non-conventional evolutionary learning paradigms [36]. While in the former, the focus is to learn and evolve the value functions associated with each state and related actions, the latter tries to evolve behaviours. There are various aspects that need to be addressed while devising multi-agent based strategies. These include homogeneous versus heterogeneous agent teams, co-operation, restricted communication, credit assignments, etc.

Communication amongst agents plays an important role for the success of any multi-agent system. Efficient communication is the foundation for effective coordination, information distribution and learning from one another. Nonetheless, an unrestricted and torrential communication essentially reduces a multi-agent system to a single agent system [42]. Further, such communication facilities are not pragmatic in real-world systems. Hence, communication among agents in a multi-agent system needs to be selective and judiciously restricted while at the same time facilitating learning and co-operation [16]. Many multi-agent frameworks disregard this objective or rather neglect communication complexities to simplify the process of information sharing and learning [3,6]. The problem becomes more complex when, in a large distributed network, asynchronous agents need to discover other similar agents with which sharing can be performed.

Multi-agent based approaches which also use the mobility of agents termed as *Mobile* agents [20,37] can be considered as a compelling paradigm to realize truly distributed yet intelligent systems. Apart from mobility, these agents have various distinguishing features such as adaptability, autonomy, on-site computation and are distributed and pervasive. They have been proved to be more efficient than the traditional point-to-point communication models [35,38]. Although the mobile agent technology is still shaping up, recent advancements in small handheld devices, miniature computers with high processing capabilities and the advent of emerging fields such as CPS and IoT have propelled the research towards the development of mobile agent based systems [8]. Since these agents migrate throughout the network to sense and process data from and at different nodes in a real distributed environment, the information they possess can vary with time. This form of information is akin to the level of experience that different persons possess while working in the same or different environments. Sharing of such experiences among the individuals could result in the growth of the overall knowledge of the group [13]. Though, researchers have used mobile agents based cooperation in a myriad of applications [22,32,35], this paper puts forward a concentrated effort on collaborative learning and information sharing among the entities of a distributed environment that utilizes the intrinsic characteristics of mobility and local execution capability of such agents.

In this paper, we present a framework for sharing intelligence and mutual learning among a set of location-unaware spatially segregated networked entities or nodes of a distributed environment; all of which have the same objective. The static agents are resident within each of the entities while the mobile ones facilitate the exchange of information and localized sharing over a network. The framework focuses on the *generic* mechanisms required to be embedded within individual agents so as to result in the overall convergence of the entire agent population, towards their common goal. The proposed framework is fully distributed in the sense that neither the agents (both static and mobile) possess any knowledge about the overall number of such agents present in the network nor do they possess any location information about other agents. Further the sharing of information among the agents is completely asynchronous and local. The major contributions of this work include:

– A multi-agent framework for distributed intelligence-sharing and learning.
– Modalities for local sharing and exchange of information.
– Dynamics for facilitating agent migration, inter-agent interactions and asynchronous executions.

The succeeding sections discuss our motivation and explore the idea of a multi-agent distributed intelligence-sharing and learning framework. A formal description of the proposed framework and the related dynamics have also been provided. The latter sections present the results along with the related discussions and conclusions arrived at.

2   Motivation

Social insect colonies provide the most natural examples of large scale multi-agent systems. These complex and self-organizing societies function based on very simple processes of information transfer between the individuals, thus providing an ideal perspective to understand the mechanisms of social learning [30]. Social insects invest considerable effort in passing on learned information to others in their group or swarm. The value of information obtained from others depends on the context [10]. Such social learning systems are often flexible enough to ensure that individuals rely on social information only when individual learning does not suffice. In an insect colony, information learned by an individual is not actually broadcasted to its peers or mediated through a supervised or centralized control. On the contrary, the information flows through local inter-actions among individuals e.g. *Trophallaxis* in bees [29] and *Tandem running* in ants [14]. It has been demonstrated that bees learn associations between floral scent and nectar rewards during trophallatic interactions, just as they would if they were to sample the flowers themselves [10]. These local interactions among the individuals within an insect colony add up to eventually alter the behaviour of the entire colony and converge their searches to the location having maximum availability of nectar.

These complex yet versatile systems of insect colonies have been a source of inspiration for many a researcher in various fields of engineering and sci-ence [43]. In the context of learning in multi-agent systems, social insect-colony based models have been of notable interest [9, 21]. In this paper, our focus is to exploit the use of social interactions among *nomadic* agents so as to facil-itate asynchronous sharing and co-operative learning among the entities of a distributed environment. Exchange of information among these *mobile* agents, populating a network of nodes, takes place locally within a node, as and when they meet other such agents within. These local interactions tend to increase the quantum of knowledge they possess. The mobile agents use this extra knowledge gained through local interactions that are spread spatially across the network, for enhancing their self-centric learning thus reducing their individual as also over-all search spaces. This learned information is thus provided to the static agents resident within the entities of the distributed environment, which evaluate the new information and provide valuable feedback for further enhancement.

For illustration, imagine a scenario wherein multiple mobile robots are trying to learn a single objective function such as solving a Rubik's cube, wrapping gifts, assembling a chair, etc. As can be observed, these tasks require a specific sequence of actions to be performed in a specified manner to achieve the desired objective. Since, there are multiple learning robots in this scenario, sharing of their individual experiences can enhance the performance of the whole system and also reduce the time the robots take to achieve the goal. However, sharing information in such a setting wherein the robots are dynamic entities is not trivial. Drawing inspiration from the social insects, one of the possible ways to share information could be by making a robot move to the vicinity of another and share information locally. The robots can gather such locally shared information over time and then use sophisticated learning tools to create a new plan of action. The robots can evaluate this new plan by executing them individually and consequently find the amount of progress they have made in achieving the goal. Repeating such a process would eventually lead all the robots towards the convergence of their shared objective. The point to be noted here is that the *mobility* of the robots contributes to the spreading of the information in the environment. However, mobility in robots is a costly affair in terms of both energy and actuation. The problem may become more critical if we think of a large network of robots wherein a few of them are only trying to learn and share a common objective.

In another scenario, assume a large Campus Area Network (CAN) populated by different kinds of devices as its nodes forming an IoT. These devices may include several PCs, projectors, air conditioners, sensors, Wi-Fi routers, printers, etc. In such a setting, imagine that a set of devices are required to find/learn to use a specific set of parameters (such as resolution (dpi), mode, paper-type, toner density, etc. in case of a set of printers) so as to optimize their life-time and utility. It is also possible that the devices need to adapt their settings based on their make and model. The simplest or naïve solution will be to package each device with an algorithm or program which always tries to figure out the optimized settings based on the user-feedback it receives on its own current settings. Assume that if the parameters are not good enough, the user changes these settings to suit her/his needs. This could be used as a feedback for the algorithm embedded within the devices. The problem with this approach is that each device (say printer) would try to solve the same problem repeatedly and hence there would be wastage of power, paper and cartridges. The life-time of the printer would also go down due to wear and tear during this laborious learning phase. A smarter way would be to share the locally learned information among the printers as in the case with mobile robots. This will not only reduce the wastage but also boost convergence time since multiple printers would be collaborating to achieve the same goal. However, it is not essential that these printers know the location of other such printers of their kind on the network. Further with no physical mobility and sophisticated programs on-board to handle communication complexities, learning optimal settings by a heterogeneous set of networked location-unaware printers, autonomously in the CAN, becomes a challenging task.

In the multi-agent based approach portrayed herein, we try to leverage the mobility based local sharing model of the mobile robots to alleviate the challenges discussed above. While a static agent manages local tasks at the physical learning entity i.e. a robot or a printer, its mobile counterparts (mobile agents) provide the much needed mobility of all the learned information. Imagine the network of robots supports a framework with all agent based functionalities as proposed in [18]. The authors in [18] describe the methodologies how a mobile agent based framework for a network of heterogeneous devices can be conceived. Let us further assume that the static agents reside within each of the networked robots. These agents manage local information and configuration of a robot such as preserving feedback, executing an action, etc. A set of mobile agents embedded with a learning algorithm suited for such an application could be released into the network of robots. These mobile agents then forage for the robots which are trying to learn within the network and facilitate the exchange of information locally with static agents hosted within each robot. Further, these agents can search the network to share their information with other such mobile agents asynchronously. When a mobile agent encounters other such agents at the various nodes in the network (not essentially the targeted robots) it exchanges information on the newly learned aspects of the solution to the problem. As a result of sharing each mobile agent comes up with a new set of actions/plan as per its learning algorithm. This new learned information is then provided to the static agent hosted within the robots which in turn executes the new plan or solution and provides its feedback to the mobile agent. The latter continues its sojourn in the network of robots after receiving the feedback. The process of learning and sharing continues till eventually all such mobile agents agree on the same set of actions indicating convergence. It may be noted that the job of collaboratively learning the optimal set of actions is achieved by the set of robots using local communications of migrating mobile agents. Similar applications such as learning optimal printer settings, finding unique genome sequences among different databases distributed across a large network, etc., can be envisaged using such multi-agent based approaches.

Although the above mentioned approach may seem trivial, it puts forward many interesting challenges that are crucial in the realization of an asynchronous and distributed intelligence-sharing and learning framework. These include:

1. The parameters both Input and Output such as number of mobile agents, learning algorithm, etc. that are essential to regulate the functioning of the framework.
2. The dynamics associated with the different parameters involved within such as agent migration, on-node execution, etc.
3. The duration for which a mobile agent should search for other mobile agents so as to share or receive information.
4. The mechanisms for the movement and exchange of learned information.
5. The formulation of conditions that will subsequently trigger the execution of the learning algorithm using the newly collected information.

In the succeeding sections, we present the proposed multi-agent based approach to model the above mentioned challenges followed by the framework in detail.

3   Proposed Framework

This section discusses the parameters, the inputs and the mechanisms required to realize the proposed multi-agent framework for distributed and asynchronous sharing of intelligence along with the formalism to model all the processes and interactions used within.

3.1   System Model

The system under consideration is modelled based on the following:

- W is an undirected connected network, where $W = (N, E)$ wherein nodes are location-unaware.
- N is a set of nodes such that $N = \{n_i | i <= C_1\}$, where C_1 is the total number of nodes in the network W, $i, C_1 \in I$ where I is a set of positive integers.
- E is a set of links such that $E = \{e_i | i <= C_2\}$, where C_2 is the total number of links in the network W, $i, C_2 \in I$.

- A is a set of static agents resident on each of the nodes such that $A = \{a_i | \forall a_i \exists n_i, n_i \in N\}$, $i \in I$.
- M is a finite set of autonomous mobile agents such that $M = \{m_i | i <= K\}$, where K is the total number of mobile agents in the system and $i, K \in I$.
- P is an application dependent user-defined learning problem whose solution (G) is to be found.
- G is a goal (objective function) provided by user.
- L is a learning algorithm provided by the user based on the underlying application which is carried by each mobile agent as its payload.
- $\eta(.)$ is a function (set of actions) that a static agent can execute in any order with or without repetitions to achieve the goal G.

A mobile agent $m_i \in M$, which is by itself an autonomous program, is capable of migrating from a node n_i to another node n_j if there exists a link e_i between (n_i, n_j), where $n_i, n_j \in N$ and $e_i \in E$ within W. Each node n_i hosts an agent framework such as [31] that is capable of managing all agent related functionalities. Each node n_i also maintains a queue (Qn_i) of mobile agents present within node n_i. The queue, Qn_i, at a node n_i is defined as

$$Qn_i = \{q_j | q_j \in M, j <= \Gamma, \Gamma < K\}, Qn_i \subset M,$$

Γ is the length of queue, $j, \Gamma \in I$.

It is apparent that if $|Qn_i| = \Gamma$, no mobile agent can enter the node n_i. Each node is a uniprocessing entity thus all the operations within a node are executed sequentially. Figure 1 shows a typical network along with the mobile and static agents, the agent framework, the learning problem and the respective queues within.

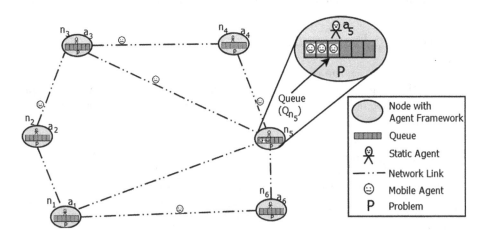

Fig. 1. A network of nodes having the agent framework, the static agent, the learning problem and the queue within. The mobile agents are shown to be either migrating from one node to another or resident within the queue of a node.

In the proposed framework, the mobile agents carry the learned information within themselves, share it with other such agents during their sojourn in the network. They also assimilate the newly gathered information to discover newer paths towards the goal G and execute the same at a node using the help of the static agents within. They collect the feedback and enrich their learned information and once again set out to collect more information from other such agents. Both, the mobile and static agents, co-exist in the network and share and execute to eventually ensure convergence.

3.2   Inter-agent Interactions

The interaction among the agents forms a crucial component of the proposed framework. As can be observed, there can be two kinds of possible agent interactions within the framework namely mobile-to-static agent interaction and mobile-to-mobile agent interaction. Since the network under consideration is distributed having location-unaware nodes, the static-to-static agent interaction is not possible under this framework.

Mobile-to-Static Agent Interaction. Mobile and static agents interact with each other in the following ways:

(a) En-queue: A mobile agent m_j can request a static agent a_i on the node n_i to execute an *en-queue* operation so as to enter Qn_i to effect its migration from another node n_k to n_i.
(b) De-queue: A mobile agent m_j can request the static agent a_i on the node n_i to execute a *de-queue* operation to enable its exit from Qn_i resulting in its migration to another node n_k from n_i.
(c) Execution of L: A mobile agent m_j at a node n_i, can execute the user-defined learning algorithm L using the computing environment provided by the static agent a_i resident at the node n_i.
(d) Execution of $\eta(.)$: A mobile agent m_j can request the static agent a_i on the node n_i to execute a set of actions and provide feedback. As $\eta(.)$ leads to the goal G (Sect. 3.1) which is dependent on the application under consideration, the executions mentioned here could imply either a set of movements for a navigating mobile robot, a set of rules for mining a large database, a set of input parameters for a control algorithm, etc.

Mobile-to-Mobile Agent Interaction. A mobile agent m_i *interacts-with* ($\otimes$) another mobile agent m_j resulting in sharing of intelligence if both the agents are present within the queue of a node n_k i.e.

$$m_i \otimes m_j \quad \textbf{iff} \quad m_i, m_j \in Qn_k, i \neq j$$

Hence, all the interactions among the mobile agents are always local and take place only *within the queue* of a node where the agents reside after migrating to a node.

3.3   Mobile Agent Migration Strategy

The mobile agents migrate within the network W using the ϵ-*Conscientious* migration strategy which is a combination of the *Random* and *Conscientious* migration strategies [33]. In the *Random* migration, a mobile agent chooses one of the neighbours of the current node at random and migrates to that node. In *Conscientious* migration strategy the mobile agent maintains a list of previously visited nodes (say V) and migrates to one that it has not visited so far. If it has visited all, it moves to the node which it has visited least recently. It can be noted that due to this migration strategy, the mobile agents tend to evenly distribute their frequency of visits at each node within the network. The *Conscientious* migration may intuitively seem better from the perspective of a single mobile agent. However, in a multiple mobile agent scenario, this strategy may lead the mobile agents to follow one another along a fixed path within the network.

In ϵ-*Conscientious* migration, a mobile agent employs the *Random* migration with a probability ϵ while it follows the *Conscientious* migration with probability $(1 - \epsilon)$. Thus, with an ϵ-*Conscientious* migration strategy, agents always try to reach out to non-visited or least visited nodes with a higher probability while reducing the drawback of a purely conscientious migration strategy.

3.4   Distributed Asynchronous Intelligence-Sharing and Learning

Let S^{m_i} be a set of shareable intelligence units (s_i) that a mobile agent $m_i \in M$ receives from the static agent $a_j \in A$ at a node $n_j \in W$. The set of shareable intelligence S^{m_i} is the learned information gathered as a result of the feedbacks obtained by m_i via the static agent a_j when it executes $\eta(.)$. Hence, the structure of individual elements $s_i \in S^{m_i}$ depends on the learning problem P of the application under consideration.

Each mobile agent m_i carries a *Bag*, B^{m_i}, (similar to the casebase of an agent as mentioned in [15]) which is a set of s_is obtained from the sets of shareable intelligence S^{m_i}, of other mobile agents as a result of sharing between m_i and the other mobile agent. Hence, B^{m_i}, which forms a part of the mobile agent's payload, can be defined as:

$$B^{m_i} = \{b_i | \exists m_j \text{ such that } b_i = s_k, s_k \in S^{m_j}, b_i \notin S^{m_i} \quad i, j, k \in I\}$$

Below we enumerate the definitions of sharing and learning within the scope of the proposed framework.

Definition 1. *A mobile agent m_i is said to have shared its intelligence with another mobile agent m_j if*

$$m_i \otimes m_j, s_i \in S^{m_i}, s_i \notin S^{m_j}, s_i \Rightarrow m_j, \quad m_i, m_j \in M, i \neq j$$

where '$\Rightarrow$' denotes that s_i is assigned to the mobile agent m_j resulting in $s_i \in B^{m_j}$.

The sharing of information amongst the mobile agents is completely asynchronous in nature. This essentially means that there is no global clock to synchronize the sharing events among the multiple mobile agents at various nodes within the framework. Thus sharing between the several mobile agents populating the network could take place concurrently at different nodes.

Definition 2. *As mentioned in Sect. 3.1, $\eta(.)$ is a function (set of actions) provided by a mobile agent m_i to a static agent a_j at a node n_j. The execution of $\eta(.)$ which is facilitated by a_j at n_j returns a new S^{m_i} which is passed on to m_i by a_j as the feedbacks. Thus $S^{m_i}_{new} = \eta(.)$. The mobile agent m_i is said to have learned new information if*

$$|S^{m_i}_{new}| > |S^{m_i}_{old}|$$

where, $S^{m_i}_{old}$ is the shareable intelligence possessed by the mobile agent m_i before the execution of $\eta(.)$ and $S^{m_i}_{new}$ is the same that the mobile agent m_i receives after this execution by the static agent a_j at node n_j.

Definition 3. *The problem P is said to have solved if a solution to goal G is found by all the mobile agents. This convergence is said to have achieved **iff***

$$\forall i, m_i \to G, m_i \in M$$

where, $\to$ denotes convergence. Hence, the main objective of the proposed framework is to ensure the convergence of all the mobile agents within the network to the common goal G using local sharing and consequent learning.

3.5 Inherent Mechanisms Within the Framework

The proposed framework provides a distributed model for sharing and learning amongst a set of location-unaware nodes in a network. In this framework we use asynchronous local sharing as a basis of information exchange by providing mobility to the learned information. Figure 2 depicts the cycle of learning that a mobile agent goes through within the proposed framework.

Each mobile agent (m_i) starts its operations initially from an *Executor* state. In this state, the mobile agent m_i residing in the queue, Q_{n_j}, of node n_j interacts with the static agent a_j also resident at the node n_j. Each mobile agent m_i is conferred with an *Execution Potential* (ξ_{m_i}) which is consumed gradually (discussed later) with every execution that the static agent a_j performs based on a request from m_i at the node n_j. The *Execution Potential* (ξ_{m_i}) restricts the mobile agent to reside at a node indefinitely. In the *Executor* state, the mobile agent m_i provides a sequence of actions derived from $\eta(.)$ to the static agent a_j as an attempt to achieve the goal G at the node n_j. This reduces the value of ξ_{m_i} based on the length of the sequence. The static agent a_j executes these sequence of actions at the node n_j. As a result of this execution, the mobile agent m_i receives feedbacks on the derived sequence of actions from the static agent a_j. This constitutes the part of the shareable-intelligence S^{m_i} gained by the mobile agent m_i at node n_j. These feedbacks also replenish ξ_{m_i} based on

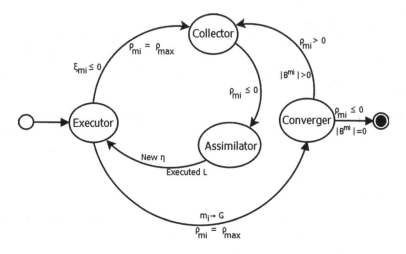

Fig. 2. The learning cycle of a mobile agent in the proposed framework

criteria discussed later. A mobile agent continues to remain in the *Executor* state until its ξ_{m_i} is exhausted. Once, $\xi_{m_i} \leq 0$, the mobile agent m_i assimilates the shareable-intelligence S^{m_i} it received and transits to the *Collector* state. In this state, the mobile agent traverses the network W, to share its shareable-intelligence with other mobile agents as well as to get shareable-intelligence from them as and when they co-exist within the same queue at a node. The mobile agent communicates locally with other mobile agents in the *Collector* state to share information. As a result of these local communications within a queue, the mobile agent m_i may receive a new set of information from the shareable-intelligence of other mobile agents. This new set of information is deposited into the *Bag*, B^{m_i} which is unique to the agent.

Every mobile agent is also empowered with a quantity termed as *Migration Resource* (ρ_{m_i}) [24] *a priori* which is also carried as its payload. The parameter ρ_{m_i} governs and regulates the duration for which the mobile agent migrates around in the network W, so as to meet and share information, in the *Collector* state. A mobile agent m_i enters the Collector state with $\rho_{m_i} = \rho_{max}$, ρ_{max} being the maximum possible value of ρ_{m_i} as defined by the user. The mobile agent m_i continues to traverse the network W, till its *Migration Resource* (ρ_{m_i}) is exhausted i.e. $\rho_{m_i} \leq 0$. This resource provides the impetus to the mobile agent to migrate within the network in search of other mobile agents having better information. The dynamics governing ρ_{m_i} have been discussed later. The structure of a mobile agent with all its components (payloads) can be seen in Fig. 3.

The mobile agent m_i transits to the *Assimilator* state from the *Collector* state as and when ρ_{m_i} degrades to a value less than or equal to zero. This is the state where the learning takes place. In this state, the mobile agent m_i uses the learning algorithm (L) which it carries as payload. It combines the information

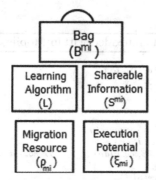

Fig. 3. Contents within the payload of a mobile agent in the proposed framework.

in S^{m_i} and B^{m_i} as the input to L and churns out a new execution plan (i.e. a new sequence of actions from $\eta(.)$) to reach the goal G at a node n_j. The generation of the new execution plan triggers the mobile agent m_i to transit to the *Executor* state.

The static agent a_j in turn executes this new plan within node n_j and provides the related feedback to the mobile agent m_i. This makes the mobile agent m_i transit back to the *Collector* state (unless of course the goal G is achieved) and the cycle in Fig. 2 continues. The value of ρ_{m_i} is again changed to ρ_{max} before it enters this cycle.

If the static agent reports that the goal G has been achieved, the mobile agent then transits to the *Converger* state with $\rho_{m_i} = \rho_{max}$. Mobile agents in this state tend to verify whether the converged goal G is an optimum or not. A mobile agent m_i in the *Converger* state migrates in the network W to find other mobile agents having a better solution than the one it has found. If the mobile agent m_i finds another mobile agent m_j having a better solution (as per the criteria defined by the user) then m_i receives the shareable-intelligence from m_j into its *Bag* B^{m_i}. This makes the mobile agent m_i to transit back to the *Collector* state and once again join the cycle. However, if ρ_{m_i} becomes less than or equal to 0 for the mobile agent m_i in the *Converger* state and B^{m_i} is still empty, it triggers the mobile agent m_i to exit the learning cycle. Hence a mobile agent m_i exits the learning cycle **iff**

$$(m_i \to G) \wedge (\rho_{m_i} \leq 0) \wedge (|B^{m_i}| = 0)$$

The overall process halts when all the mobile agents exit the learning cycle through the *Converger* state. The Algorithm 1 depicts all the functions of a mobile agent in the proposed framework as described above.

3.6   Dynamics Within the Framework

As mentioned, both ρ_{m_i} and ξ_{m_i} act as fuel for migration and on-node execution on part of the mobile agents, respectively. The dynamics that regulates the values ρ_{m_i} and ξ_{m_i} are discussed below.

Algorithm 1. Algorithm embedded within Mobile Agents

	Input	: η, L, G, P, ρ_{max}	
	Output	: Convergence path to G	
	Initialization: $B^{m_i} = \{\}$, $S^{m_i} = \{\}$		
1	**while** *Not converged to G* **do**		
2	enter_node(n_j);	// Executor State //	
3	**while** $\xi_{m_i} > 0$ **do**		
4	$Exec\_Plan$ = action_sequence(η);		
5	initialize(ξ_{m_i}, $Exec\_Plan$);		
6	to_StaticAgent(a_j, $Exec\_Plan$) ;	// Mobile to Static Agent Interaction	
7	receive_feedback();		
8	retrieve_intelligence(S^{m_i});		
9	update_potential(ξ_{m_i});		
10	**end**		
11	**if** *G is not reached* **then**		
12	$\rho_{m_i} = \rho_{max}$;	// Collector State //	
13	**while** $\rho_{m_i} > 0$ **do**		
14	migrate_next_node();		
15	en_queue(Q_{n_j});		
16	ρ_{m_i} = migration_penalty(ρ_{m_i}, B^{m_i});		
17	**if** *information request received* **then**		
18	share_intelligence(S^{m_i});	// Mobile to Mobile Agent Interaction	
19	**end**		
20	**if** *end_of_queue* **then**		
21	find_shareable_agent();		
22	**if** *agent is available* **then**		
23	$W_{old} = \Pi_{m_i}(B^{m_i})$;		
24	b_i = get_shareable_intelligence();	// Mobile to Mobile Agent Interaction	
25	$B^{m_i} = B^{m_i} \cup b_i$;		
26	$W_{new} = \Pi_{m_i}(B^{m_i})$;		
27	ρ_{m_i} = migration_reward(ρ_{m_i}, W_{old}, W_{new});		
28	**end**		
29	**end**		
30	**end**		
31	**end**		
32	**else if** *G is reached* **then**		
33	$\rho_{m_i} = \rho_{max}$;	// Converger State //	
34	**while** $\rho_{m_i} > 0$ **do**		
35	migrate_next_node();		
36	en_queue(Q_{n_j});		
37	ρ_{m_i} = migration_penalty(ρ_{m_i}, B^{m_i});		
38	validate_convergence(G);		
39	**if** *G is non-optimal* **then**		
40	Jump to *Collector* State ;		
41	**end**		
42	**end**		
43	**end**		
44	**if** $G \wedge (\rho_{m_i} \leq 0) \wedge (|B^{m_i}| = 0)$ **then**		
45	exit() ;	// Convergence	
46	**end**		
	// Assimilator State //		
47	run_learning_algorithm(L, B^{m_i}, S^{m_i});		
48	get_new_sequence(η);		
49	**end**		

Dynamics of ρ_{m_i}. We assume that each piece-wise intelligence $b_i \in B^{m_i}$ gained in the learning exercise by the mobile agent m_i in the *Collector* state, to achieve the goal G, has a value or weight associated to it. This weight is synonymous to the profit gained or loss incurred, as the case may be, in using this piece of intelligence to advance towards the goal G. Let $\phi(.)$ be the function which returns this weight for each piece-wise intelligence $b_i \in B^{m_i}$. Hence, the net weight (Π_{m_i}) of the *Bag*, B^{m_i}, is calculated as:

$$\Pi_{m_i} = \sum_i \phi(b_i), \forall b_i \in B^{m_i} \tag{1}$$

The weight function $\phi(.)$ depends on the underlying application and can be designed based on a knowledge-sharing model as proposed in [13].

A mobile agent always enters the *Collector* state with $\rho_{m_i} = \rho_{max}$, where ρ_{max} is the maximum possible value of ρ_{m_i} conferred on it *a priori*. A migration penalty is incurred on ρ_{m_i} whenever a mobile agent m_i in the *Collector* state moves to a new node.

The value of ρ_{m_i} at the new node is computed as:

$$\rho_{m_i}(x_{n+1}) = \begin{cases} \rho_{m_i}(x_n)e^{-\Pi_{m_i}} & \text{, if } \Pi_{m_i} > 0 \\ \rho_{m_i}(x_n)(1 - \frac{1}{\rho_{max}}) & \text{, otherwise} \end{cases} \tag{2}$$

where, x_n denotes the nth instance.

As can be observed in the above equation, the value of ρ_{m_i} degrades exponentially with increase in the weight of the *Bag* B^{m_i}. As more nuggets of information populate the *Bag* B^{m_i} of a mobile agent m_i due to local communication with other mobile agents, the payload of the mobile agent m_i increases and makes its movement across the network W sluggish. The migration resource ρ_{m_i}, however, decreases accordingly and inhibits the mobile agent m_i from further migration when $\rho_{m_i} \leq 0$. This triggers the mobile agent to enter *Assimilator* state wherein it generates a new plan using the algorithm L. On the contrary, if the *Bag* B^{m_i} is empty, the mobile agent's payload is lighter and ρ_{m_i} is high, forcing it to explore for fresh information across the network W by migrating to other nodes in search of other mobile agents that can provide the same.

While the above equation tends to reduce ρ_{m_i} of an agent due to migrations, it is also *recharged* whenever a mobile agent in the *Collector* state receives information from another mobile agent as a result of local sharing. This also means that whenever there is an increase in weight Π_{m_i} of the *Bag* within a mobile agent m_i, the value of ρ_{m_i} increases empowering it to travel further into the network W in spite of its sluggishness caused by the heavy *Bag* B^{m_i}. The value of ρ_{m_i} when a mobile agent m_i receives and accumulates new information into its *Bag* B^{m_i} is calculated as:

$$\rho_{m_i}(x_{n+1}) = \rho_{m_i}(x_n) + c\frac{\delta}{\Pi_{m_i}}\rho_{max} \tag{3}$$

where,

$$\delta = \Pi_{m_i}^{after\_sharing} - \Pi_{m_i}^{before\_sharing} \qquad (4)$$

c is a constant and $c > 0$.

Dynamics of ξ_{m_i}. The *Execution Potential* (ξ_{m_i}) of a mobile agent decreases linearly with the execution of every action within the action-sequence when the mobile agent m_i is in the *Executor* state. As mentioned earlier, the action-sequence is derived from $\eta(.)$ which is performed by the static agent.

$$\xi_{m_i}(x_{n+1}) = \xi_{m_i}(x_n) - k_1, \; k_1 > 0 \qquad (5)$$

However, when the mobile agent receives a positive feedback (defined by the user for the specific problem whose solutions are being learnt using the algorithm L) from the static agent as a result of executing an action, the *Execution Potential* (ξ_{m_i}) is topped up as:

$$\xi_{m_i}(x_{n+1}) = \xi_{m_i}(x_n) + k_2, \; k_2 \geq 0 \qquad (6)$$

k_1 and k_2 are constants.

The increase in ξ_{m_i} allows for further exploration into the search space of the problem that has not been achieved as a result of sharing with other mobile agents. It thus aids the generation of new piecewise information and subsequent enhancement of the set S^{m_i}.

4 Implementation

Since the work reported herein exploits the simultaneous or concurrent executions of multiple agents running and sharing in parallel at different locations (nodes) within a network, experiments if conducted on an inherently sequential simulator would grossly undermine the strength and ability of the proposed framework. Hence, we have implemented the whole framework using the *Typhon* [31] platform over a Local Area Network (LAN). *Typhon* provides for mobile agent programming on real networks and runs over the proprietary *Chimera* Static Agent Platform that is shipped along with LPA WIN-PROLOG (http://www.lpa.co.uk/). Hence, along with static agents, *Typhon* provides all the mobile agent related functionalities such as migration, execution, cloning, payload carrying capability, etc. One instantiation of a *Typhon* based platform over *Chimera* acts as a node in the network. Several such instantiations were used to realize various overlay networks of different sizes varying from 10 nodes to 50 nodes over the LAN. Further, to evaluate the robustness of the proposed framework, experiments were conducted using a dynamic network of 50 nodes emulating a mobile computing environment. As the experimental test-bed was completely implemented over a LAN, the results gathered also involve the real-time states (such as processing speed, network conditions, etc.) of different machines used in the experiments. One separate dedicated computer served as a log server to record events such as sharing, executions, etc., at the various nodes.

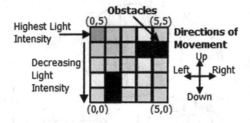

Fig. 4. The schematic of a 5 × 5 *Maze-World*. The top left corner in the *Maze-World* is the location of highest light intensity and formed the Destination location. The difference in colour variation of the cells show the change in light intensity as we move away from the destination. The black cells depict obstacles. The four directions of movement in the *Maze-World* are shown separately (Color figure online).

The events were time-stamped using the local time at the log server as and when the relevant pieces of information were received from the individual nodes in the network.

4.1   Terms Used in the Implementation

In this section, a glossary of various terms used in the implementation has been provided for better clarity. The meaning of terms used are as follows:

- Virtual robot: This is a simulated robot which has been tasked to learn a sequence of actions to reach a specified goal.
- *Maze-World*: This is an $n \times n$ grid structure with each cell of the grid having a set of sensor vectors.
- Sensor Vector (SV): This is a set of sensor values perceived by the Virtual Robot in a cell within the *Maze-World*.
- Static agents: These agents have the responsibility of executing operations on the virtual robots and to interact with the mobile agents.
- Mobile agents: They act as the carrier of learned information from one virtual robot to another.
- Node: The node comprises the *Typhon* Agent Platform, Queue for mobile agents, *Maze-World* and the virtual robot managed by a static agent.
- Network: This is the inter-connection of nodes which facilitates migration of mobile agents from one node to another.

4.2   The Distributed Learning Problem (P)

For evaluation of the proposed framework, a problem involving a set of virtual robots assigned with the task to learn a path from a fixed source (S_S - lowest light intensity) to a fixed destination (S_D - highest light intensity) in a *Maze-World* was used. The visualization of the *Maze-World* is shown in Fig. 4 and the corresponding networked-framework is depicted in Fig. 5. As can be seen,

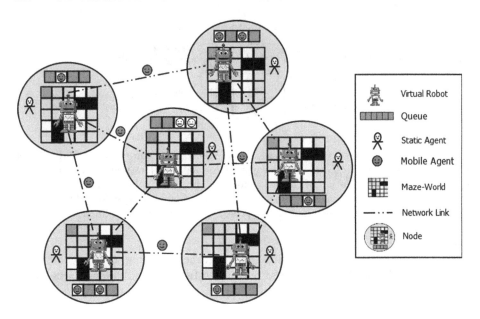

Fig. 5. An approximate visualization of the virtual robot network along with the *Maze-World*.

the *Maze-World* is available within each virtual robot locally. The set of virtual robots form the nodes of the distributed network. A static agent was also stationed at each node to manage all the functions of the virtual robot on the *Maze-World* available within the node. As can be observed, the *Maze-World* and virtual robot replace the problem component P and the node in Fig. 1 respectively. The network W is formed by the set of virtual robots situated within the nodes with the *Maze-World* and static agent. The mobile agents use this distributed network to disseminate the learned information received from the virtual robots via the static agents.

4.3 Specifications of the Virtual Robot

Each virtual robot is equipped with three virtual sensors viz. a *Light* sensor, an *Obstacle* sensor and a *Direction* sensor. The value of the *Light* sensor increases as the virtual robot moves towards the destination and is highest at the destination. Its value is inversely proportional to the distance between the destination and the current location of the virtual robot in the *Maze-World*. Obstacles may also populate the *Maze-World*. A virtual robot can sense these obstacles using its Obstacle sensor which returns only a binary value viz. *true* or *false*. The sensor returns a *true* if there is an obstacle in the direction of the virtual robot's heading or when it encounters the boundary of the *Maze-World*; else it returns a *false*. A virtual robot cannot cross any obstacle or the boundary of the *Maze-World*. Further, the value of the *Light* sensor becomes 0 if the value of *Obstacle* sensor is *true.* The *Direction* sensor returns the direction of movement of the virtual robot in the *Maze-World*. There can be four possible values of the *Direction* sensor viz. *Up (positive Y-axis)*, *Down (negative Y-axis)*, *Right (positive X-axis)* and *Left (negative X-axis)*. Hence, a virtual robot is not allowed to move diagonally from one location to another.

The virtual robot can perform five actions within the *Maze-World*. These actions are as follows:

- *Move_Forward:* Execution of this task makes the virtual robot to change its location in the direction of its heading one step at a time. For example, if the current location of the virtual robot is (2, 3) and its current heading is *Up*, then the new location of the virtual robot would be (2, 4) after the execution of the action *Move_forward*.
- *Move_backward:* Execution of this action changes the location of the virtual robot in the diametrically opposite direction of its heading without changing the current heading of the virtual robot. For example, if the current location of the virtual robot is (3, 3) and the current heading direction is *Left* then after executing the *Move_backward* action, the new location of the agent would become (4, 3) and the heading will remain *Left*.
- *Turn_Left:* This action changes the heading of the virtual robot 90° towards the clockwise direction of its current heading. The location of the virtual robot within the *Maze-World* is not affected by this action.
- *Turn_Right:* This action changes the heading of the virtual robot 90° towards the anticlockwise direction of its current heading. The location of the virtual robot within the *Maze-World* is not affected by this action.
- *Turn_Back:* This action changes the heading of the virtual robot to 180° of its current heading without changing the location of the virtual robot in the *Maze-World*.

4.4   Embedding the Problem P in the Framework

All the virtual robots have been embedded with the sensor vectors (SVs) of the source (S_S), and the destination (S_D) within the *Maze-World* using which they come to know about the source and the destination. The goal G for the virtual robot is to find a series of SV transitions using the available actions to reach the destination viz. S_D from the source S_S. The network of virtual robots uses the proposed framework to realize the mentioned objective.

$\phi(.)$ is defined as the difference of the norm between the SV before and after an action is executed i.e.

$$\phi(X) = ||SV_o|| - ||SV_i|| \tag{7}$$

where, X is an action, SV_i is the sensor vector before the execution of the action and SV_o is the sensor vector after the execution of the action.

The S^{m_i} in this case is the set of SVs with $\phi(X) > 0$. The structure of a piecewise unit of intelligence $s_i \in S^{m_i}$ together with an example s_is sequence are shown in Fig. 6(a) and (b).

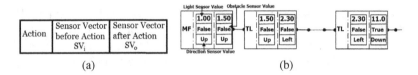

Fig. 6. (a) Structure of a piecewise intelligence $s_i \in S^{m_i}$ that a mobile agent m_i shares with other mobile agents (b) An example of s_is forming a sequence

4.5   The Learning Algorithm (L)

In our implementation, we have used a greedy approach to construct the sequence of actions towards the destination. When the mobile agents are in the *Assimilator* state of the learning cycle (Fig. 2), they perform an incremental search within their *Bags* to find the associated transitions of SVs to reach the destination S_D. This search explores various combinations of SVs available within the *Bag*, B^{m_i} and S^{m_i}, of a mobile agent m_i and outputs the largest possible sequence of actions based on the SV transitions. Two piece-wise intelligence $b_i, b_j \in B^{m_i}$ such that $i \neq j$, can be connected to form a link if $SV_o \in b_i$ (i.e. $SV_o^{b_i}$) is equal to $SV_i \in b_j$ (i.e. $SV_i^{b_j}$) i.e.

$$SV_o^{b_i} = SV_i^{b_j}$$

Thus, SV transitions or b_is result in a tree with either the start vector S_S (the SV of the starting point in the *Maze-World*) or an SV closest to S_S (based on the *cosine* distance among the SVs) at the root of the tree. Other b_is within the *Bag*, B^{m_i}, form the node of the tree. The algorithm L uses *Depth-First-Search* (*DFS*) and outputs the branch having the highest length. In case, two branches have the same length, it outputs the branch with highest weight, Π_{m_i} (calculated using Eq. 1).

After assembling the sequence of actions towards the goal, G, the mobile agent transits to the *Executor* state. As mentioned earlier, in *Executor* state the static agent at a node executes the sequence of actions on the virtual robot thus evaluating the sequence of actions obtained using L and records the fresh SVs. If the execution of all the actions is over and the *Execution Potential*, ξ_{m_i}, is still greater than zero, then the static agent selects an action at random out of the set of actions ($\eta(.)$) and executes them. This allows for the self-discovery on the part of the agents which can be shared with others.

4.6   Complexity Analysis Within the *Maze-World*

Let us assume that the size of the *Maze-World* is $n \times n$. One may note that there could be more than one SV possible within a cell in the *Maze-World*. For example in the present *Maze-World* there are four different SVs, one for each direction within a cell. If S is the number of SVs possible within each cell of the *Maze-World*, the total possible SVs within the *Maze-World* would be Sn^2. Let

T be the number of actions that can be performed and l be the length of the sequence of actions to be executed to reach the destination (S_D). If we consider that the starting point for all virtual robots is fixed (S_S) within the *Maze-World*, then the total search space would reduce to $O(T^l)$.

If we relax our assumption that the length, l, of the sequence of actions is known, then for a single agent, the total time complexity to reach to the destination location (S_D) in the worst-case would be:

$$\beta := O\left(\sum_{l=1}^{Sn^2-1} T^l\right)$$

This gives us the upper bound on the time complexity of the search space using a single virtual robot positioned at a fixed location within the *Maze-World*.

Further, let us assume that α number of virtual robots at a fixed location within the *Maze-World* are trying to move towards the given destination. In the best case when no redundant executions are performed due to one-to-all sharing amongst virtual robots, the total time complexity of exploring the *Maze-World* by α virtual robots is of the order of β/α plus the overheads incurred in sharing.

Let the time taken for sharing intelligence between two virtual robots be τ. Apparently, the sharing of information involves two virtual robots (one who provides the information and the other one who receives it) at a time. Hence, the total number of sharing events required to disseminate the intelligence within each virtual robot among the remaining virtual robots would be $^{\alpha}C_2$. It may be noted that the virtual robots can share intelligence concurrently. If α is even, then the number of concurrent sharing events possible is $\alpha/2$. Thus the total time required to share the intelligence among the α virtual robots would be $\frac{(2\tau \; ^{\alpha}C_2)}{\alpha}$.

In case if α is odd, this time complexity would become $\left(2\frac{^{\alpha-1}C_2}{\alpha-1} + \alpha - 1\right)\tau$.

Hence, the total time complexity to reach the destination in the best case for α virtual robot (α being even) would be:

$$\gamma := O\left(\frac{\beta + (2\tau \; ^{\alpha}C_2)}{\alpha}\right)$$

This forms the lower bound on the time complexity of the overall search space. Let θ be the time complexity of solving the problem P using the proposed framework. Since, in our proposed framework $\alpha > 1$ along with localized sharing using the concept of mobility of learned information, intuitively θ will be bounded as: $\gamma < \theta << \beta$

It may be noted that these bounds are not so tight on θ, yet they give us an approximation of the reduction in the search space and time complexity when the proposed sharing framework is used.

5   Results

Experiments were carried out on various networks of virtual robots with different populations of mobile agents to learn a sequence of actions (i.e. the goal G in this case) to reach the destination. Each experiment was performed at least 10 times to counter any stochastic influence. The time required to complete each of the experiments varied between 300 to 3000 s. The average of 10 runs has been portrayed in the results. The values of various parameters used for the experiments are:

$\epsilon = 0.2$, $\rho_{max} = 10$, $n \times n$ (Size of maze) $= 50 \times 50$, Start Location (S_S) $= (50, 0)$, Destination Location (S_D) $= (0, 50)$

The mobile agents were placed arbitrarily at different nodes within the network of virtual robots during the start of each of the experiments. All the nodes in the network of virtual robots were connected in the form of a mesh.

For experimentation, *Typhon* [31] based networks with sizes varying from 10 nodes to 50 nodes were created. The densities (D) of the mobile agents within a network, which can be defined as the ratio of number of agents to the number of nodes in the network, were varied from $D = 0.1$ to 0.5 on each of the networks and all the executions were logged. The experiments involved the problem of finding the sequence of actions required to traverse from the starting location S_S to the destination location S_D within the *Maze-World* with no obstacles. To vary the problem setting, we have also experimented on a 50-node network with the *Maze-World* having contiguous obstacles occupying co-ordinate locations $(40, 0)$ to $(40, 40)$.

Two important factors that are crucial to verify the effectiveness of the proposed framework are - the size of the network (i.e. number of nodes in the network) and the number of mobile agents involved. While the former tests the scalability of the framework, the latter can ensure a faster convergence. Hence, as a performance yardstick, we varied the density (D) of mobile agents in networks of different sizes and recorded the average number of executions that the virtual robots took to converge to the goal G. By number of executions of virtual robots, we mean the number of times each mobile agent entered the *Executor* state (Sect. 3.5). Thus, the average number of executions is the ratio of the total number of executions (until the convergence of all the mobile agents in the network) to the total number of mobile agents.

5.1   Converged Sequences of Actions

The graphs in Fig. 7(a) and (b) depict the final converged paths taken by the virtual robots with 5 mobile agents in the 50-node network. Figure 7(a) shows the converged paths of five virtual robots without sharing i.e. when the proposed sharing framework was not used. While, a total of 5 mobile agents ($D = 0.1$) in the 50-node network were used with the proposed framework for the graph in Fig. 7(b). As can be observed in Fig. 7(a), when the proposed sharing framework is not used by the virtual robots, all the virtual robots discover different sequences of actions (paths) to the destination S_D for the goal G. However, all

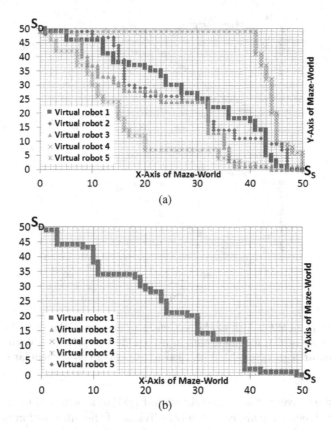

Fig. 7. Final converged sequences of five agents in the *Maze-World* from the Source at (50, 0) to the Destination at (0, 50) in a 50-node network - (a) Without Sharing (Paths are distinct) (b) With Sharing (Paths are highly overlapped).

the virtual robots converge to the same sequence of actions when they share information with one another. These graphs not only show the impact of sharing but also reveal the expected performance of the proposed framework. The goal G, assigned to the virtual robots, was to find a sequence of actions (as discussed in Sect. 3.5) that facilitate their movement from a source at (50,0) to a destination at (0,50). It can be seen clearly that sharing of information definitely helps these virtual robots achieve their goals in lesser time with fewer number of executions. From the logs, we have found that the average number of executions was 195 when the virtual robots did not share information whereas it was a mere 92 when sharing was embedded using the proposed framework. It should be noted that all the virtual robots are required to find their path individually. The results indicate that the proposed framework allows mobile agents to search in different directions (possibilities) and the best amongst them is taken up by all the virtual robots to beeline towards the goal. Similar trends were observed in all other cases considered for the experimentation.

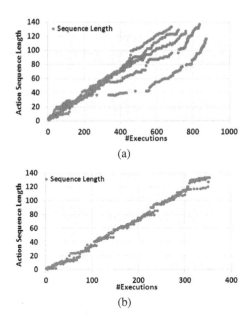

Fig. 8. Variations in the length of the sequences of actions of the virtual robots with executions in case of five mobile agents in a 50-node network - (a) Without Sharing (Mostly dissimilar sequences) (b) With Sharing (Converge on the similar sequences).

Further the converged sequence shown in Fig. 7(b) is achieved by five different mobile agents when they are completely oblivious of the information about other mobile agents within the network along with any knowledge pertaining to the network itself (i.e. the number of virtual robots). Hence, the proposed framework augments the algorithm L and facilitates the sharing of information amongst the virtual robots in a truly distributed sense and also yields better performance.

5.2   Length of the Sequences of Actions

The graphs in Fig. 8(a) and (b) depict the variations in the length of the sequence of actions with respect to the executions on part of the virtual robots with five mobile agents whose converged sequences are shown in the graphs in Fig. 7(a) and (b) respectively. The differences in the lengths of the sequences discovered by the five virtual robots against their executions can be observed in Fig. 8(a) when they are not sharing any intelligence amongst each-other. As can be observed in the case when the mobile agents shared intelligence (Fig. 8(b)), the lengths of the sequences of all the virtual robots are spread evenly and confined along a common line even before convergence when they are in the process of discovering the paths which is the period when execution increases. This illustrates that with no sharing of intelligence, the virtual robots search egocentrically and explore to find their individual solutions. Whereas the virtual robots are able to effectively constrict their search space towards the goal by sharing their piecewise knowledge

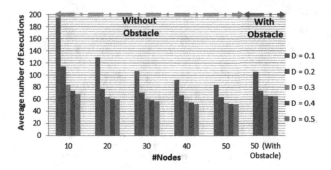

Fig. 9. Variations in the average number of executions to converge to a path from the source to the destination within a *Maze-World* of size 50×50 by all the mobile agents for different densities. The graph is plotted by taking the average of 10 runs in each case on *Typhon* based networks with no obstacles in the *Maze-World* for the first five sets. The last set is one with obstacles stretching from (40,0) to (40,40) in the *Maze-World* (Color figure online).

using the dynamics of the proposed framework. Each virtual robot thus tries to solve the sub-problems of a common agenda. It can also be observed that the total number of executions performed by all the five mobile agents taken together which is around 350 is less than half of the total number of executions taken together of all the non-sharing virtual robots (around 900). Hence, the results portrayed herein show that the proposed framework makes the virtual robots agree on a common solution and share the best available information amongst each other using the mobility of the learned information. The graphs in Fig. 8 show one instance of the results obtained. Similar trends were observed when the number of nodes and mobile agents were varied in all other cases considered for experimentation.

5.3   Average Number of Executions

Figure 9 depicts the average number of executions required to learn a path from source to the destination within the *Maze-World* of size 50×50 with varying number of nodes and mobile agents.

As can be observed from the graph in Fig. 9, the average number of executions (until all the mobile agents converge to a path to S_D) reduces monotonically with increase in the density of the mobile agents. Further, the trend remains the same as the size of the network (number of nodes in the network) grows. One can also observe that as the density becomes high (50 % of the total number of nodes), no significant difference in the average number of executions across different network sizes is observed. It may be noted that in the real-world, the execution of a series of actions is a costly affair both in terms of energy and time. Also, $D = 0.1$ in a 10-node network depicts a scenario when there is no sharing since there is only 1 mobile agent in the network. The average number of executions in this case is 195. While the average number of execution reduces to almost half (110 executions)

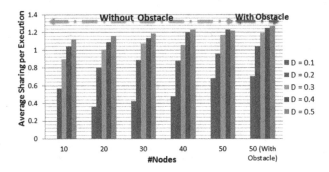

Fig. 10. Average number of sharing events per execution with different densities. The graph is plotted by taking the average of 10 runs in each case on *Typhon* based networks with no obstacles in the *Maze-World* for the first five sets. The last set is one with obstacles stretching from (40,0) to (40,40) in the *Maze-World* (Color figure online).

with $D = 0.2$ in case of 10 nodes. This depicts the huge gain in terms of average number of executions when the mobile agents are sharing information against the case when they are not doing so. Further, the average numbers of executions are higher in case of 50 nodes when there was obstacles within the *Maze-World* against the no-obstacle case. This shows that the complexity of the problem does affect the performance though the trend of executions remains the same. It may also be noted that the reduction in the average number of executions is high when the density of mobile agents varies from $D = 0.1$ to 0.2 in all the cases considered for experimentation. This reduction slows down from 0.2 onwards and almost becomes asymptotic. Moreover, increasing the density above 0.5 is not advisable as the population of mobile agents would clutter the network and entail more communication overheads [19]. Hence, the results clearly show that the proposed framework can effectively bring down the number of executions required to achieve the goal G in a fully distributed and asynchronous manner.

5.4 Average Number of Sharing Events per Execution

The graph in Fig. 10 depicts the average number of times the sharing was performed per execution with varying densities and sizes of the network. The average number of sharing events per execution is calculated as the ratio of the average of the total number of sharing events performed in 10 runs of each experiment to the average of the total number of executions in 10 runs. The cases considered are same as described in the previous section. As can be observed there is an increase in the average number of sharing events per execution with increase in the density of mobile agents in each of the networks of different sizes. Further, as observed in the previous case, the trend remains the same across different network sizes and problem setting. Apparently, the average number of sharing events per execution is 0 in case of $D = 0.1$ in the network with 10 nodes. The graph also reveals the fact that an initial increase in the density of mobile agents aids to expand the search in multiple directions which increases the amount of sharing. However, a high density of mobile agents within the network causes redundancy in the search space and hence creates less amount of shareable intelligence amongst each other. Further, the graph also shows that a change in the problem setting (*Maze-World* with obstacles) does not alter the trend or the behaviour of the sharing among the mobile agents. It can also be observed that the amount of sharing increases when the problem becomes more complex (*Maze-World* with obstacles) because of the increase in size of the search space.

It may be noted that there is a marginal difference (< 0.05) in the average sharing per execution between $D = 0.4$ and $D = 0.5$ across all cases. These are possibly the best operating densities of the number of mobile agents in the proposed framework to effectively use distributed asynchronous sharing in the current problem setting of the *Maze-World*. Though a lower number of mobile agents could eventually find the solution (taking more number of executions and less sharing), an optimum number of mobile agents in the network could hasten the process while also effectively utilizing the network resources. A mechanism to dynamically vary the density of mobile agents based on the current size of the network as reported in [19] could aid the performance of the proposed framework.

5.5   Distinct Movements of Virtual Robots Within the *Maze-World*

The graphs in Fig. 11(a) and (b) depict the movement of all the virtual robots within the *Maze-World* of size 50×50 for $D = 0.1$ in a 50-node network without and with obstacles. The final converged sequences of actions (G) of all the virtual robots are also depicted in the graphs. These two cases of *Maze-World* exhibit different levels of complexities for the virtual robots to reach the destination location (50, 0) from the starting location (0, 50). When there is no obstacle within the *Maze-World*, the obvious way to traverse the *Maze-World* to reach the destination is to move diagonally while in the presence of an obstacle the best choice is to follow the path alongside the obstacle towards the destination. In case of an obstacle, the virtual robots need to exert more energy in terms of the number of executions required to explore the path alongside the obstacle as compared to a diagonal path in the absence of the same. Insertion of multiple obstacles along the path also gave similar results. As can be observed from the graphs, the final converged sequence of actions (G) in both without and with the obstacles, for all the virtual robots is the best possible intersection of the sequences available to reach the destination location at (50,0). Since the learning algorithm discussed in Sect. 4.5 does not ensure the optimality in the movement of virtual robots towards their destination, such an effect evolves as a result of the distributed asynchronous sharing of the piecewise intelligence available within each mobile agent which in turn circulates the knowledge of the global best among all the virtual robots. Sharing of information on part of the mobile agents seems to bring back the virtual robots that drift away from the optimal path resulting in a drastic reduction in the number of otherwise futile executions. It thus motivates the entire population of virtual robots to pursue a common and possibly more optimal path in an *orderly* and *unified* manner.

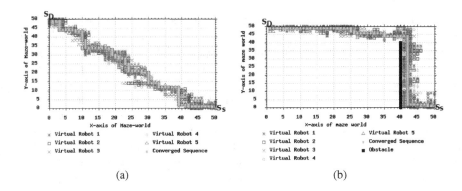

Fig. 11. Movements of all five virtual robots with sharing from the start of the experiment until convergence - (a) with $D = 0.1$ in a 50-node network with no obstacles in the *Maze-World* and (b) with $D = 0.1$ in a 50-node network with an obstacle stretching from location $(40,0)$ to $(40, 40)$ in the *Maze-World*.

5.6   Frequency of Sharing

Figure 12 shows the change in number of sharing events versus time along with the associated linear trendline for $D = 0.5$ in a 50-node network. It may be noted that the X-axis indicates the time-stamps recorded by the log server (as mentioned in Sect. 4). As can be seen, the linear trendline drawn based on the frequency of sharing events has a negative slope. Since the mobile agents try different combinations of actions on to the virtual robots initially they tend to discover diverse piecewise solutions causing the sharing events to be high. The number of sharing events decreases gradually as the virtual robots move towards the goal. This is because the initial high level of sharing causes the mobile agents to converge closer to a common path causing lesser diversity of information within their respective bags resulting in a gradual reduction in sharing. A similar

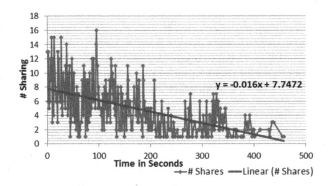

Fig. 12. Number of sharing events among 25 agents in a 50-node network ($D = 0.5$) against time in seconds.

trend was observed in all other cases of varying nodes and mobile agents used in the experimentation.

Sharing thus seems to constrain the mobile agents to search within a common and narrower search space. This could lead the virtual robots to follow a non-optimal path (local optima). One alternative to circumvent this could be to embed a different learning algorithm within a few mobile agents so as to force them explore the search space in a different manner ensuring diversity of the contents in the various *Bags* for a longer time. This increase in diversity of the piecewise solutions carried within the individual *Bags* of the mobile agents will consequently increase sharing and aid in an exit path from possible local optima.

5.7   Performance in a Dynamic Network

Figure 13 depicts the performance of the proposed distributed intelligence-sharing and learning framework in a 50-node dynamic network. The dynamic network was created using a variant of the Erdős-Rényi $G(n, p)$ model [11]. Initially a mesh network of 50 nodes was established over a LAN. Each node within the network was provisioned to break their connections with any of their current neighbours with a probability of 0.3. A node was allowed to make a new connection with any other node in the network with a probability of 0.5. Each node exercises the event of altering their connections after a time interval randomly chosen between 10 and 20 s in real time. Thus, the virtual robots within the dynamic network mimicked the movement or mobility of actual mobile robots just as in a mobile ad-hoc network by breaking and making new connections with other robots as they move physically. From the logs, the minimum degree of a node (d) was found to be 0 while the maximum degree was found to be 27. When a virtual robot becomes isolated ($d = 0$), then all the mobile agents within that node become *dormant* and wait till the virtual robot reconnects. Hence, the total number of nodes in the connected network keeps on varying with time.

As can be observed from Fig. 13(a), the average number of executions (until all the mobile agents converge to G) reduces with increase in the density of the mobile agents within the 50-node dynamic network. This follows the same trend as observed in case of the static networks discussed earlier. Further as shown in Fig. 13(b), the average number of sharing events per execution increases with increase in the number of mobile agents within the network. The trend of the graph in this case also remains similar to the previously cited cases of static networks. As can be seen from the graph, there is an initial boost in the average number of sharing events per execution in case of 5 mobile agents to 10 mobile agents while the same slows down as we move from 10 mobile agents to 25 mobile agents. As mentioned earlier, this is due to the fact that an initial increase in the density of agents aids to expand the search in multiple directions resulting in an increase in the overall sharing. As the density of mobile agents increases, redundancy in information discovered by the mobile agents increases thereby lowering the quantum of shareable intelligence.

It may thus be observed that the proposed framework is suited to networks wherein the *nodes are mobile*. This emphasizes the robustness and flexibility of

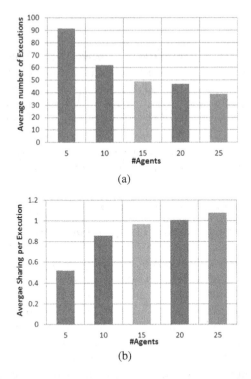

Fig. 13. Performance in the *Typhon* based 50-node dynamic network of virtual robots with varying densities from $D = 0.1$ (5 agents) to 0.5 (25 agents) (a) Average number of executions to converge to a path from the source to the destination within a *Maze-World* of size 50×50 by all the mobile agents (b) Average number of sharing events per execution.

the proposed framework for intelligence-sharing and learning in distributed environments. The results substantiate the viability of using the proposed framework in distributed mobile computing environments.

6   Discussions

The proposed framework opens numerous avenues for testing and verifying a gamut of learning algorithms over distributed, decentralized and asynchronous environments. Convergence is hastened due to faster access to newly weaned information gained from sharing on part of the mobile entities in the network. The multi-agent paradigm coupled with this mobility of learned information, facilitates collaborative learning among a set of location-unaware entities or nodes in a large distributed, decentralized and asynchronous system. The intrinsic flexibility of agent based technology used by the framework could also allow learning of multiple goals by the various entities in a network. Thus, in an N-node network, I nodes could be trying to find a solution to a problem P_1, J nodes could be doing the same for another problem P_2, while the rest $(N - I - J)$ try to solve P_3. These problems may need to be solved concurrently or could even be made sequential using techniques cited in [24, 25]. In such a case, the N-node network would be hosting heterogeneous sets of mobile agents which use the proposed framework to solve their respective problems. Heterogeneity in this case refers to the difference in the parameters and the learning algorithms used by the three sets of mobile agents attempting to solve the problems P_1, P_2 and P_3 respectively as mentioned above. Given a heterogeneous set of algorithms say L_1, L_2 and L_3 to be used to solve a particular problem P, such a set up of the framework could also be used to evolve the best algorithm. In this case, three sets of mobile agents could be primed to solve the problem P using an algorithm respectively within the proposed framework. Consequent to this, each set of mobile agents would evolve three solutions S_1, S_2 and S_3 based on the algorithms L_1, L_2 and L_3 respectively. In order to find the efficacy of the solutions S_i and find the best among them, an Idiotypic network [23] based mechanism such as the one proposed in [26] could be used. The mechanism proposed in [26] could allow the N-node network to automatically evolve the best solution out of S_1, S_2 and S_3.

In addition, if it so happens that an algorithm L_1 performs optimally during the initial phase of learning but unfortunately deteriorates in its performance at later stages while another algorithm L_2 behaves just the reverse way then running them together within the framework could be meaningful. While the set of mobile agents using L_1 will clearly dominate in performance initially, the other set of mobile agents using L_2 will use this learned information and enhance the convergence at a rapid pace. The heterogeneous set of mobile agents using different learning algorithms can thus co-operate and lead to a better solution using this framework. Since the performance of different algorithms could vary depending on the problem at hand such a hybridizing of algorithms could pave ways to newer and more efficient mechanisms.

The framework could also be enhanced by supplementing it with cloning of the best performing sets of mobile agents as described in [19]. Accordingly, the cloning of mobile agents would not only enhance concurrent processing by increasing the population of the best performing agents, but also restrict the lesser performing ones from consuming precious system resources such as bandwidth and computation times. Modifying and emulating variants of population based algorithms such as Evolutionary Computing and Genetic Algorithms (GA) including Island models [7] can also be performed using this framework.

As discussed earlier, the design methodology of the proposed framework is inspired by the interactions in social insect colonies and draws concepts from the domain of Evolutionary Computing and Genetic Algorithms (GA). Modifying and emulating variants of such population based algorithms can also be performed using this framework. If we consider each mobile agent carrying the learned information as a candidate solution, then this framework can be viewed as a real-time formulation of a distributed GA. The support for mobility of candidate solutions in this framework also allows the real implementation of Island models of parallel GA [7]. Such models have been proposed as a distributed implementation but have hardly been used, possibly due to the non-existence of a convincing platform to realize them. If we look for the analogies between the proposed framework and GA, it may be observed that the Migration Resource, ρ_{mi}, mildly mimics the crossover operator used in a GA. It also makes candidate solutions to be nomadic and forces their movement from one island population to another. Further, the Execution Potential, ξ_{m_i}, partially mimics the mutation operator used in a GA by forcing the system to try out new solutions. Given a set of actions, it makes the system try out new permutations and combinations of these actions. The feedback provided by the static agent at the nodes participating in the learning exercise can be attributed to the fitness of the solutions in a GA. Similar analogies can also be derived for other population based bio-inspired algorithms such as Particle Swarm optimization (PSO) [27]. The proposed framework can thus be effectively utilized as a potential tool for intelligence-sharing and learning in large distributed systems in a variety of scenarios.

7   Conclusions

Distributed intelligence-sharing and learning in multi-agent systems, open up a wide variety of applications, that range from sensor beds to smart cities [5]. Mobile agent technology can play a crucial role in optimizing the use of local computing resources and distributed decision making in highly complex and scalable systems. This paper attempts to highlight the advantages of distributed intelligence-sharing and learning using a set of mobile agents. The *emulation* experiments presented in this paper provide valid proof-of-concept for the same. The model of intelligence-sharing and learning proposed herein could be used as a framework to realize distributed and continuously evolving systems which in turn could exhibit emergent behaviours. A variety of user-defined learning algorithms could also be used concurrently by different sets of agents within this framework. Further, since the number of mobile agents required is unknown, the initial discovery could commence with a moderate number of mobile agents, some of which may be allowed to clone over a period of time based on a performance measure so as to hasten the process of convergence. Excessive cloning can however lead to cluttering within the network which could be prevented using an appropriate controller such as the one described in [19].

It may be noted that the proposed framework makes use of information available locally (within a node) to achieve a global objective. This is akin to the functioning of densely populated insect colonies such as ants, honeybees, etc. [10]. The asynchronous sharing and consequent learning observed in the proposed framework could provide insights to the manner in which these swarms interact and converge towards a goal. It further opens up avenues to explore applications in the domain of asynchronous robotic swarms wherein every execution consumes precious energy stored within their batteries. Such a system could thus enhance the capabilities of the robotic swarm several manifolds. It is possible to conjecture each robot in a swarm as a mobile agent and then embed the whole mechanism of local sharing and learning to constitute an intelligent mobile ad hoc network of robots. The framework described herein can also be used in many Internet of Things (IoT) and Cyber-Physical Systems (CPS) based applications. Some of the examples of such applications can be - energy management in smart buildings [44] wherein a set of mobile agents could be used to learn a schedule of energy consumption by various networked devices, intelligent water distribution systems [17] wherein such agents could share the information of a distributed network of various water reservoirs and optimize water distribution. Other applications could be the learning of the occupancy patterns of large buildings [34] for regulating HVAC systems and distributed intelligent traffic monitoring and management systems [4] wherein the agents can share the information of heavy traffic routes and learn traffic regulation rules to cater to different times of the day thus evolving optimized route schedules. In industry automation such agents can share the information of workloads on different remotely located machines and learn an optimized schedule for job allocations on-the-fly while in intelligent warehouse management systems these agents can facilitate sharing of information of the items placed in various smart racks and shelves to come up with the better and optimal strategies to improve logistics and placement of goods, etc. Use of a set of mobile agents, empowered with the sharing and learning mechanisms, in all these scenarios is bound to instil and enhance intelligence in such networked environments.

Acknowledgements. The first author would like to acknowledge Tata Consultancy Services (TCS) and Ministry of Human Resource Development, Govt. of India for the support rendered during the research reported in this paper.

References

1. Alonso, E.: Multi-agent learning. Auton. Agent. Multi-agent Syst. **15**(1), 3–4 (2007). http://dx.doi.org/10.1007/s10458-007-0019-1
2. Atzori, L., Iera, A., Morabito, G.: The internet of things: a survey. Comput. Netw. **54**(15), 2787–2805 (2010). http://www.sciencedirect.com/science/article/pii/S1389128610001568
3. Berenji, H., Vengerov, D.: Advantages of cooperation between reinforcement learning agents in difficult stochastic problems. In: The Ninth IEEE International Conference on Fuzzy Systems, 2000, FUZZ IEEE 2000, vol. 2, pp. 871–876 (2000)
4. Bode, M., Jha, S.S., Nair, S.B.: A mobile agent based autonomous partial green corridor discovery and maintenance mechanism for emergency services amidst urban traffic. In: Proceedings of the First International Conference on IoT in Urban Space, URB-IOT 2014, pp. 13–18. ICST (Institute for Computer Sciences, Social-Informatics and Telecommunications Engineering), Brussels (2014). http://dx.doi.org/10.4108/icst.urb-iot.2014.257297
5. Brenna, M., Falvo, M.C., Foiadelli, F., Martirano, L., Massaro, F., Poli, D., Vaccaro, A.: Challenges in energy systems for the smart-cities of the future. In: 2012 IEEE International on Energy Conference and Exhibition (ENERGYCON), pp. 755–762, September 2012

6. Busoniu, L., Babuska, R., De Schutter, B.: A comprehensive survey of multiagent reinforcement learning. IEEE Trans. Syst. Man Cybern. Part C Appl. Rev. **38**(2), 156–172 (2008)
7. Cantú-Paz, E.: A survey of parallel genetic algorithms. Calculateurs Paralleles, Reseaux et Systems Repartis **10**(2), 141–171 (1998)
8. Cao, J., Das, S.K.: Mobile Agents in Networking and Distributed Computing, vol. 3. Wiley, Hoboken (2012)
9. Cicirello, V., Smith, S.: Wasp-like agents for distributed factory coordination. Auton. Agent. Multi-agent Syst. **8**(3), 237–266 (2004). http://dx.doi.org/10.1023/B%3AAGNT.0000018807.12771.60
10. Dukas, R.: Insect social learning. In: Moore, M.D.B. (ed.) Encyclopedia of Animal Behavior, pp. 176–179. Academic Press, Oxford (2010). http://www.sciencedirect.com/science/article/pii/B9780080453378000589
11. Erdős, P., Rényi, A.: On the evolution of random graphs. Publ. Math. Inst. Hungar. Acad. Sci **5**, 17–61 (1960)
12. Ferber, J.: Multi-agent Systems: An Introduction to Distributed Artificial Intelligence, 1st edn. Addison-Wesley Longman Publishing Co., Inc., Boston (1999)
13. Fisch, D., Jnicke, M., Kalkowski, E., Sick, B.: Learning from others: exchange of classification rules in intelligent distributed systems. Artif. Intell. **187188**, 90–114 (2012). http://www.sciencedirect.com/science/article/pii/S0004370212000410
14. Franks, N.R., Richardson, T.: Teaching in tandem-running ants. Nature **439**(7073), 153–153 (2006)
15. Garland, A., Alterman, R.: Autonomous agents that learn to better coordinate. Auton. Agent. Multi-agent Syst. **8**(3), 267–301 (2004). http://dx.doi.org/10.1023/B%3AAGNT.0000018808.95119.9e
16. Ghavamzadeh, M., Mahadevan, S., Makar, R.: Hierarchical multi-agent reinforcement learning. Auton. Agent. Multi-agent Syst. **13**(2), 197–229 (2006). http://dx.doi.org/10.1007/s10458-006-7035-4
17. Giannetti, L., Maturana, F.P., Discenzo, F.M.: Agent-based control of a municipal water system. In: Pěchouček, M., Petta, P., Varga, L.Z. (eds.) CEEMAS 2005. LNCS (LNAI), vol. 3690, pp. 500–510. Springer, Heidelberg (2005). http://dx.doi.org/10.1007/11559221_50
18. Godfrey, W.W., Jha, S.S., Nair, S.B.: On a mobile agent framework for an internet of things. In: 2013 International Conference on Communication Systems and Network Technologies (CSNT), pp. 345–350, April 2013
19. Godfrey, W.W., Jha, S.S., Nair, S.B.: On stigmergically controlling a population of heterogeneous mobile agents using cloning resource. In: Nguyen, N.T. (ed.) TCCI XIV 2014. LNCS, vol. 8615, pp. 49–70. Springer, Heidelberg (2014). http://dx.doi.org/10.1007/978-3-662-44509-9_3
20. Harrison, C.G., Chess, D.M., Kershenbaum, A.: Mobile Agents: Are They a Good Idea?. IBM TJ Watson Research Center Yorktown Heights, New York (1995)
21. Holland, O.E.: Multiagent systems: lessons from social insects and collective robotics. In: The 1996 AAAI Spring Symposium on Adaptation, Coevolution and Learning in Multiagent Systems, pp. 57–62 (1996)
22. Ilarri, S., Mena, E., Illarramendi, A.: Using cooperative mobile agents to monitor distributed and dynamic environments. Inf. Sci. **178**(9), 2105–2127 (2008). http://www.sciencedirect.com/science/article/pii/S002002550700583X
23. Jerne, N.K.: Towards a network theory of the immune system. Annales d'immunologie **125**, 373–389 (1974)

24. Jha, S.S., Godfrey, W.W., Nair, S.B.: Stigmergy-based synchronization of a sequence of tasks in a network of asynchronous nodes. Cybern. Syst. **45**(5), 373–406 (2014). http://dx.doi.org/10.1080/01969722.2014.917235

25. Jha, S.S., Nair, S.B.: Orchestrating the sequential execution of tasks by a heterogeneous set of asynchronous mobile agents. In: Müller, J.P., Weyrich, M., Bazzan, A.L.C. (eds.) MATES 2014. LNCS, vol. 8732, pp. 103–120. Springer, Heidelberg (2014)

26. Jha, S.S., Shrivastava, K., Nair, S.B.: On emulating real-world distributed intelligence using mobile agent based localized idiotypic networks. In: Prasath, R., Kathirvalavakumar, T. (eds.) MIKE 2013. LNCS, vol. 8284, pp. 487–498. Springer, Heidelberg (2013)

27. Kennedy, J.: Particle swarm optimization. In: Gass, S.I., Fu, M.C. (eds.) Encyclopedia of Machine Learning, pp. 760–766. Springer, New York (2010)

28. Konstantinidis, A., Yang, K., Zhang, Q., Zeinalipour-Yazti, D.: A multiobjective evolutionary algorithm for the deployment and power assignment problem in wireless sensor networks. Comput. Netw. **54**(6), 960–976 (2010). http://www.sciencedirect.com/science/article/pii/S1389128609002679, new Network Paradigms

29. Korst, P., Velthuis, H.: The nature of trophallaxis in honeybees. Insectes Soc. **29**(2), 209–221 (1982). http://dx.doi.org/10.1007/BF02228753

30. Leadbeater, E., Chittka, L.: Social learning in insects from miniature brains to consensus building. Curr. Biol. **17**(16), R703–R713 (2007)

31. Matani, J., Nair, S.B.: *Typhon* - a mobile agents framework for real world emulation in prolog. In: Sombattheera, C., Agarwal, A., Udgata, S.K., Lavangnananda, K. (eds.) MIWAI 2011. LNCS, vol. 7080, pp. 261–273. Springer, Heidelberg (2011). http://dx.doi.org/10.1007/978-3-642-25725-4_23

32. Miller, K., Mansingh, G.: Towards a distributed mobile agent decision support system for optimal patient drug prescription. In: 2013 Third International Conference on Innovative Computing Technology (INTECH), pp. 233–238, August 2013

33. Minar, N., Kramer, K., Maes, P.: Cooperating mobile agents for dynamic network routing. In: Hayzelden, A., Bigham, J. (eds.) Software Agents for Future Communication Systems, pp. 287–304. Springer, Berlin Heidelberg (1999). http://dx.doi.org/10.1007/978-3-642-58418-3_12

34. Oldewurtel, F., Sturzenegger, D., Morari, M.: Importance of occupancy information for building climate control. Appl. Energy **101**, 521–532 (2013). http://www.sciencedirect.com/science/article/pii/S0306261912004564, sustainable Development of Energy, Water and Environment Systems

35. Outtagarts, A.: Mobile agent-based applications: a survey. Int. J. Comput. Sci. Netw. Secur. **9**(11), 331–339 (2009)

36. Panait, L., Luke, S.: Cooperative multi-agent learning: the state of the art. Auton. Agent. Multi-agent Syst. **11**(3), 387–434 (2005). http://dx.doi.org/10.1007/s10458-005-2631-2

37. Papaioannou, T., Edwards, J.: Building agile systems with mobile code. Auton. Agent. Multi-agent Syst. **4**(4), 293–310 (2001). http://dx.doi.org/10.1023/A%3A1012758908423

38. Queloz, P.A., Villazn, A.: Composition of services with mobile code. Auton. Agent. Multi-agent Syst. **4**(4), 311–337 (2001). http://dx.doi.org/10.1023/A%3A1012711025262

39. Rajkumar, R.R., Lee, I., Sha, L., Stankovic, J.: Cyber-physical systems: the next computing revolution. In: Proceedings of the 47th Design Automation Conference, DAC 2010, pp. 731–736. ACM, New York (2010). http://doi.acm.org/10.1145/1837274.1837461

40. Ren, W., Beard, R., Atkins, E.: A survey of consensus problems in multi-agent coordination. In: American Control Conference, 2005, Proceedings of the 2005, vol. 3, pp. 1859–1864, June 2005

41. Santos, A., Delbem, A., London, J.B.A., Bretas, N.: Node-depth encoding and multiobjective evolutionary algorithm applied to large-scale distribution system reconfiguration. IEEE Trans. Power Syst. **25**(3), 1254–1265 (2010)

42. Stone, P., Veloso, M.: Multiagent systems: a survey from a machine learning perspective. Auton. Robots **8**(3), 345–383 (2000). http://dx.doi.org/10.1023/A%3A1008942012299

43. Van Dyke Parunak, H.: "Go to the ant": engineering principles from natural multi-agent systems. Ann. Oper. Res. **75**, 69–101 (1997). http://dx.doi.org/10.1023/A%3A1018980001403

44. Zhao, P., Suryanarayanan, S., Simoes, M.: An energy management system for building structures using a multi-agent decision-making control methodology. IEEE Trans. Ind. Appl. **49**(1), 322–330 (2013)

Author Index

Printed in the United States
By Bookmasters

Printed in the United States
By Bookmasters